IMAGES
of America

FALK
COMPANY LUMBER TOWN OF THE AMERICAN WEST

On the Cover: The company town of Falk, California, was home to a venturesome and often audacious group of men and women who, with the support of an evolving cast of tools that ranged from ox to train and from axe to chain saw, cut, milled, and shipped redwood products from Humboldt County's Elk River valley to all parts of the world. In operation from 1884 to 1937, Noah Falk's Elk River Mill and Lumber Company is now an interpretive site in the Headwaters Forest Reserve, which is managed by the Bureau of Land Management. (Courtesy of Larry Hill.)

IMAGES
of America

FALK
COMPANY LUMBER TOWN OF THE AMERICAN WEST

Julie Clark
Foreword by Jon Humboldt Gates

ARCADIA
PUBLISHING

ISBN 978-1-4671-2975-6

Published by Arcadia Publishing
Charleston, South Carolina

Printed in the United States of America

Library of Congress Control Number: 2018944627

For all general information, please contact Arcadia Publishing:
Telephone 843-853-2070
Fax 843-853-0044
E-mail sales@arcadiapublishing.com
For customer service and orders:
Toll-Free 1-888-313-2665

Visit us on the Internet at www.arcadiapublishing.com

This book is dedicated to Paul Mazzucchi, who allowed me to see the past through his eyes.

Contents

Foreword

The company mill town of Falk and its people flourished for many decades. Like other small mill towns of the era, Falk fell into ruin during the Great Depression. Most of its 400 town residents moved away, and Falk became a ghost town in the 1950s. I discovered it in 1964 while hiking the forests of the Upper Elk River. In those days, the old mill's bandsaw and tools still stood in place. There was a general store, train trestles, a cookhouse, the post office, a few cars, the gas station, scores of homes, and the railroad roundhouse. My grandfather and great-grandfather had lived in Falk a century before. My family still knew some of the longtime upper Elk River families whose ancestors had worked at Falk. Over the years, I recorded the stories of people who grew up there. Those stories are told in *Falk's Claim: The Life and Death of a Redwood Lumber Town*. Little exists today of the original town. Falk was again cast into the spotlight during a historic confrontation with Maxxam Corp's Pacific Lumber Company when thousands of protesters vowed to stop the logging of old-growth redwoods in the Headwaters Forest east of Falk. Today, one of the entrances to the Headwaters Reserve passes right through where the town once stood. Falk's locomotive roundhouse, now the Headwaters Education Center, is the only surviving building of what once was Falk, California.

—Jon Humboldt Gates

ACKNOWLEDGMENTS

Sometimes in life, just the right person steps through the door when you need help at just the right time. Kurt Angersbach came to the Bureau of Land Management (BLM) Arcata Field Office shortly after the book proposal was accepted. Kurt helped me research, digitize photographs, and write captions, and above all, was there as a constant encouraging force. He has volunteered many hours in the completion of this book and deserves much kudos and respect for this project.

I came to study Falk in my early years as a park ranger for the BLM in 1999. My key reference was *Falk's Claim: The Life and Death of a Redwood Lumber Town*, written by Jon Humboldt Gates. He and I became friends, and he "passed the Falk torch" by donating his photograph collection. Much of the images and stories in here are possible because of him. A large majority of the Falk community photographs were donated by Dale Callihan, whose relatives were Falk residents. He and his wife, Jacquie, were generous in their photographs and family history.

I was fortunate to interview several Falk residents before they passed, primarily Paul Mazzucchi (1908–2006), who worked for Elk River the 1920s as a "riggin man." Without him, I would not have had the opportunity to look into the past through his eyes. Thanks go to Lewis Olson, Carolyn Bugenig, Wayne Miller, and Dorothy Fleckenstein. Thanks also go to Pete Johnston, whose father, Riggs, was the forester during the ghost town era and constantly encouraged my Falk research. Pete also wrote the chapter introduction to the "Ghost Town" chapter.

Many thanks are extended to Carly Marino at the Humboldt Room, Humboldt State University, Jim Garrison from the Humboldt County Historical Society, and Ben Brown from the Clarke Historical Museum. They were of great assistance in allowing me to use high-quality photographs and research material.

Without the constant support of the BLM, my coworkers, and my supervisors, this would have never happened. They have always given me the room to grow as an employee in my passions at Headwaters and the history that proceeded it. My husband, Cris Johnson, and father, Paul Denton, have been my personal cheerleaders during this time and had my back at all times, just like family does.

INTRODUCTION

Walking through the present-day Headwaters Forest Reserve, along the Elk River Trail, one gets a sense of timelessness as he or she looks at the big-leaf maple trees, the second-growth redwoods, and the Sitka spruce all growing next to the Elk River. Now a critical habitat for endangered species, the reserve is a place to enjoy the silence from a hard workweek, connect with nature, and slow down with family and friends. Long ago, a different community existed on the very soil that is now protected and in public hands. The trees and the foliage that have grown since the last Falk building was razed are now a thriving ecosystem, putting the past in perspective. Mother Nature does supersede, especially in Humboldt County with its rainforest climate. Without the interpretive signs along the trail, one would have no idea that hundreds of people worked, played, and dreamed in the company town of Falk. In 1884, Noah Falk, with the help of investors—his brothers Jonas and Elijah—bought virgin old-growth redwoods and built the Elk River Mill and Lumber Company, along with the company town of Falk.

Company towns in America during this time were not limited to logging, but included mining and other resource extraction. Without the option of commuting or transportation, workers basically built their own homes and gave 10 hours of their time in exchange for housing, food, and amenities. This was life for many people who worked in company towns from the 1880s to the 1930s. Westward expansion and the industrial age heeded the call for a solid workforce ready to manufacture in the cities, mine coal in the Appalachians, and build and operate mills and rails in the isolated areas of the west. Most company town owners and investors owned all the land, houses, and the business. They provided housing, food, luxuries, and a place to work for those who were willing. The company town, therefore, became especially significant in the successful expansion of mining and lumber industries of the west and the manufacturing coal industry in the east.

Noah Falk came to San Francisco in 1854 from Ohio. Eventually, he landed in Humboldt County in the early 1860s and became a successful lumber business entrepreneur. His prominence in Arcata, owning two lumber mills, gave him the accolades and clout as a mill and logging expert. In the late 1880s, old-growth redwood trees were being sold at $2.50 an acre. Investors from far and wide, illegal and legal, were jumping at the opportunity to log the ancient trees. It was no surprise when Noah was approached by investors who had just bought much of the South Fork Elk River watershed. He sent for his brothers Elijah and Jonas from Ohio, who knew just how to build a mill and manage it. Many company town owners at this time exploited their workers, as they did the land, but Noah Falk was fair and worked right alongside them, giving them a living wage of 25¢ an hour.

Many immigrants and various ethnicities were attracted to the company town atmosphere. The newly arrived immigrant would find family or fellow countrymen already working in the same community; they did not have to concern themselves with finding a place to live near their place of work. Lumber towns like Falk particularly attracted men from Scandinavian countries because they already had the skills from their homelands. Logging towns in the Pacific Northwest actually requested people from Finland, Norway, and Sweden. The exchange between lumber town owners and the Nordic immigrants is described as a win-win situation. Immigrants could not find steady work in their home countries; lumber town owners needed a workforce with prior skills. Once the immigrants got to this country, they saved their money and then sent for more family members and friends. The company gained a strong workforce, and in turn, immigrants started a new life in America. Unfortunately, in some towns, including Falk, the diversity of ethnicities was coupled with segregation, lower wages, unskilled work, and no benefits. Although the residents did their best to accept one another, the caste system was evident and not surprising in the little town of 400 people. The rise of technology contributed to creating the company towns in the industrial age, and it also brought its demise in the 1920s and 1930s with the wide use of the automobile. The isolated towns of the West were becoming more populated and closer to conveniences, and opportunities for commuting were plentiful. Limited resources contributed to the closing of the mills and mines. The Depression also effectively closed many towns due to their high operating cost and limited demand for their products. Corporations bought out company towns and viewed the housing of employees frustrating and complicated. Home mortgages and loans were also made more available to new home buyers during this time, and people simply moved out into a place of their own. In the case of Falk, the town was closed in 1930, but many residents remained. The mill and logging equipment were in dire need of updates. Winfield Wrigley, Falk superintendent, was able to secure a million-dollar loan and update the Dolbeer logging engine to tractors. The mill opened once again in 1936 but closed permanently in 1937. Ultimately, it did not take long for the vegetation to grow back and take the space of the buildings. For the next 40 years, it became a lure for bottle hunters and curious onlookers until it was razed in 1979. The space that held a mill town for 50 years is now a transitioning forest with remnants and memories of Falk resurfacing from time to time.

Much of this research was inspired by my work as the Headwaters park ranger who, from the beginning, was mesmerized by this redwood forest, its recent history, and the stories. Over the years, countless numbers of people contributed stories, photographs, and artifacts. The town's existence was relatively short in the bigger picture of Humboldt County history, but I have forever found that Falk holds a special place in many people's hearts: a time when life was simpler, when one could make a living wage on 25¢ an hour, marry the girl next door, and live off homegrown fruits and vegetables—and the community was a person's life.

One

Elk River Mill and Lumber Company

Noah Falk did not know when he left Ohio at the age of 18 that he would become one of the most influential lumbermen in Humboldt County. He would own three mills in Arcata by the time he was 48 years old and have just over 30 years of experience in and outside the mill. Much of the land he had owned in Arcata had been logged and milled by 1882. He was well regarded and trusted by the logging community, so it was no surprise in 1882 when Frank Taylor and Judge Stafford approached him for his expertise in logging hundreds of acres in the South Fork Elk River watershed (present-day Headwaters Forest Reserve). A new company was formed called the Elk River Mill and Lumber Company, with Noah Falk, Stafford, and San Franciscan J.C Hawley as equal shareholders. Noah turned to his brother Elijah for constructing and designing the Elk River Mill and his other brother Jonas to manage the mill. Between the 1880s and the 1930s, America saw a rise in the development of company towns in resource extraction businesses such as timber and mining. Due to the isolation of the Elk River Mill, commuting was not an option, and the town of Falk was born in 1884. Many of the workers had families, and the town swelled to about 200 residents. Although isolated, the inhabitants had comforts of home, with a cookhouse, dance hall, general store, and post office, along with the Jones Prairie School for the children. Many of the workers were immigrants from Sweden, Norway, Ireland, and Nova Scotia. This was also a period of innovation in logging and milling equipment. Much of the new technology transitioned from manual methods, such as using oxen to move logs, to the invention of the Dolbeer steam donkey. Milling equipment also saw big changes when replacing the circular saws with bandsaws made cutting large redwood logs more efficient.

Noah H. Falk was born in Pennsylvania in 1837 and lived in Hancock County, Ohio, until 1854. At the age of 18, he headed out west for the California Gold Rush. He soon discovered that money was in "redwood gold" when he took a position as a sawyer in Albion, Mendocino County. In 1867, he moved to Humboldt County and originally planned to open a bakery but instead began working for William Carson as a lead sawyer. (Courtesy of Jon Humboldt Gates.)

By the 1880s, Noah Falk was well known in the lumber community as a businessman with practical experience. Investors Judge Cyrus G. Stafford and J.C. Hawley of San Francisco propositioned Noah Falk to be a stakeholder, and they formed the Elk River Mill and Lumber Company. Noah's brother Elijah began construction of the mill site, located just outside of Eureka on the South Fork Elk River. By 1882, the mill was under construction, and the new company town of Falk was formed in 1884. Pictured from left to right are Herb Christie, Jim Copeland, Noah Falk, and an unidentified chopper. (Courtesy of Jon Humboldt Gates.)

During the late 1800s, Noah Falk and Isaac Minor partnered and operated two mills in Arcata. The Dolly Varden Mill, located just south of Camp Curtis, produced 20,000 board feet of lumber in 1872, and the Jolly Giant Mill (pictured) was located northwest of Humboldt State University. The Jolly Giant Mill in 1878 produced 45,000 board feet of lumber per day. With the experience and the investors, Noah was able to buy the Elk River timberland, which would later become the Elk River Mill and Lumber Company and the town of Falk. (Courtesy of Humboldt County Collections [HCC] Photos, Humboldt State University Library.)

Charles Falk (1869–1957), son of Noah and Nancy Falk, succeeded his uncle Jonas as the Elk River Mill foreman when he was 30 years old in 1900. He had 12 years previous experience in all operations including the mill, blacksmith shop, and the woods. He retired after 30 years at the age of 60 and moved to Arcata, into the Falk Victorian on Thirteenth and H Streets, where present-day Wildberries is. (Courtesy of Jon Humboldt Gates.)

In July 1890, six years after the construction of the Elk River Mill, the sawdust that coated the surfaces of the mill was ignited by a single spark. Immediately, the area was engulfed in flames. Noah Falk and other men created a line from the mill to the river, scooping up five-gallon oilcans full of water but to no avail. The mill and 100,000 board feet of lumber burned, making the total loss over $25,000. Noah and his partners were not insured and had to decide whether or not to rebuild the mill. Factors such as the investments they had already made and the uncut timber they owned weighed on their decision to rebuild. Noah brought out used equipment from his Arcata mill, and recruited his brother Elijah to rebuild, and they were open for business and producing lumber by the fall of 1890. (Courtesy of Jon Humboldt Gates.)

Company towns and logging camps were scattered all throughout Humboldt County during the early 1900s with names like Crannell, Samoa, Scotia, Falk, Dolbeer and Carson Camp, and Big Lagoon. They all had attributes in common: a cookhouse and cabins for the workers, which were portable and were moved on skids by train; a railroad running through each of the towns; and men and women who worked 10 hour days or longer. These towns were ideal for the owners, who needed an available workforce, and the workers, who needed stability and community. Although many of the company town structures are gone, there are houses and the last vestiges at Samoa, including a popular tourist destination, the Samoa Cookhouse. The last official company town in Humboldt County was Scotia. In 2011, residents voted to be independent of the New York hedge fund company that had obtained Scotia in bankruptcy proceedings. (Courtesy of HCC Photos, Humboldt State University Library.)

At the same time the Elk River Mill was being built, the Bucksport & Elk River Railroad (B&ER) was under construction. The tracks ran from Bucksport (Humboldt Bay) to the 9,000 acres of the Elk River Mill and Lumber Company. It became readily apparent that Noah's lumber company could not financially support both the mill and the seven and a half miles of railroad, and they sold their interest to Dolbeer & Carson Lumber Company and the California Redwood Company. Dolbeer & Carson Lumber Company dominated the logging industry during this time. Carson's wealth was reflected in his mansion, towering over Carson's mill and dock (pictured). The steam train owned by Dolbeer & Carson and California Redwood Co. would haul Falk's milled lumber to Humboldt Bay. In 1902, Dolbeer & Carson made an agreement with the B&ER to allow trackage rights for its trains to haul all redwood products from Elk River to the bay. B&ER sold its locomotives, the Bucksport and Trinidad for scrap in 1934. The railroad tracks were removed in 1953, but visitors can still see their remnants adjacent to Elk River Road. (Courtesy of Peter E. Palmquist Photographs, Yale Collection of Western Americana, Beinecke Rare Book and Manuscript Library.)

The 1893 Chicago World's Fair was known as the World's Columbian Exposition and was the fair to end all fairs. One in four Americans took the journey out to Illinois to see inventors from around the world show off countless new gadgets, including primitive fax machines, new telephones, an electric railway, neon lights, and forestry products. It was no surprise that a piece of Falk history found its way to the fair. The redwood plank pictured was taken by boat to San Francisco and moved by train to Detroit, costing over $3,000 for transportation. The plank was shaped by hand using a broadax during the winter of 1889, 1890, and 1891 by the Elk River Mill and Lumber Company. It was cut from a stump near Falk measuring 12 by 16 feet and 5 inches thick. The plank was named "Mammoth Redwood Plank" by a Detroit varnishing company, Berry Brothers. The company used the plank as a marketing campaign for its varnishing product and showcased in the Forestry and Horticulture building at the fair. It was renowned as being the largest piece of finished lumber on the grounds. (Both, courtesy of Jon Humboldt Gates.)

Revolutionary advances in resource extraction prompted equally revolutionary advances in resource conservation. Public interest groups formed to protect the redwoods, including the Sempervirens Club in 1900 and Save the Redwoods League in 1918. Public policy at the state and federal levels also addressed forest conservation, with the establishment of the California Redwood Park in 1902, the US Forest Service in 1905, and the National Park Service in 1916. The tools and techniques that had helped to make the oxen obsolete also triggered changes in public and governmental perceptions regarding logging operations in the redwoods. (Courtesy of the Palmquist Collection, Humboldt State University Library.)

This image of the Falk woods from 1886 depicts the brief period of overlap between oxen teams and steam power in California's redwood forests. More powerful steam donkey engines, introduced in the 1890s, coupled with the all-weather capabilities of railroads meant that the use of draft animals in logging operations would effectively come to a close by the 1920s. (Courtesy of Jon Humboldt Gates.)

Falk had two main cookhouses, also known as dinner sheds. There was one for the mill workers (pictured) and one for the loggers that moved as the work moved up the canyon. The cookhouse culture and language had its own words like *spuds* (potatoes), *sinkers* (doughnuts), and *hen fruit* (eggs). The staff was often women who rose at 4:00 a.m., prepped, cooked, served, and then washed all the heavy dishes by hand. The cook was often like a captain of the ship; it was the cook's kitchen, and not even the camp boss had a say. In order to have the stamina for the 10-hour shifts, loggers ate enormous quantities of food—an average of 8,000 calories a day. Rigging men, fallers, and buckers are said to have burned 8 to 12 calories per minute, with heavy gear and foul weather adding to the caloric burn. (Both, courtesy of Jon Humboldt Gates.)

This panoramic view of Falk taken in 1907 depicts the town at its peak just after the 1906 San Francisco earthquake. The demand for redwood was at a surplus during this time, and rebuilding the biggest city to the south was an urgency. At this time, the Elk River Mill and Lumber Company had almost 100 employees and the town about 200 residents. Located in the tight Elk River canyon, the railroad paralleled Elk River as well as most of the town structures. The structure to the left

near the railroad tracks is the general store and the gas station. In the center of the photograph are stacked milled lumber piles and bachelor cabins, and to the right are the millpond and dance hall. Thirty years later, the town would shut its doors for good, becoming a ghost town for forty years and eventually returning back to forest ground. (Courtesy of Kristi Wrigley.)

Beginning in 1881, the logging industry was revolutionized by the invention of the Dolbeer steam donkey. Named after Humboldt local citizen and engineer John Dolbeer, this upright wood-burning steam generator powered the cable run spool, which pulled logs to the nearest railroad tracks for transport. (Courtesy of Jon Humboldt Gates.)

The size and sheer abundance of the trees in the redwood forests of the American West encouraged the development of new logging tools. Here, a Falk woods crew strikes a pose with a steam donkey engine. The steam donkey, a vertical boiler with drum reels mounted on skids to aid portability, used manila rope (and later steel cable) to move logs from the woods to river or railroad pickup points. (Courtesy of Jon Humboldt Gates.)

This dam-like structure was built shortly after the construction of the mill in 1884. It held the South Fork Elk River at bay, which created a storage facility, officially called the millpond, for the logs until they could be moved to the cold deck adjacent to the mill. Most mills in Humboldt County used this method for storing logs before they were milled. (Courtesy of Jon Humboldt Gates.)

These early 1900 loggers and mill workers stand next to the Elk River Lumber Company train tracks near the Falk mill. Notice one of the workers peaking his head out the window in a playful manner. It was not all hard work, but a camaraderie grew between the men from many parts of the world, such as Sweden, Finland, Italy, and Ireland. Company towns such as Falk became a portal for immigrating to the United States. (Courtesy of Jon Humboldt Gates.)

Built in 1898, this buttress was used to support the millpond dam. Note the scale of the workers in relation to the size of the logs. Photography in the 1890s often relied on subjects maintaining a pose in order to avoid blurring the image. Imagine the effort required to maneuver these massive logs into place especially given that tools of the day included oxen, horses, manila rope, steam donkeys, and dynamite. Compare the heroic scale of this image with the famous image of Rockefeller Center under construction in 1932 in which workers pose on a beam dangling into empty space over the New York City skyline. An important difference between the images is the genuine remoteness of the Elk River Mill and the company town of Falk and the urban, heavily populated site of the New York City skyscraper. Both images show the capacity of humans to create seemingly impossible structures with limited tools and larger-than-life results. (Courtesy of the Palmquist Collection, Humboldt State University Library.)

The first significant structure to be removed from the Falk woods was one of the first to have been built there by Noah and Elijah Falk. In the early 1940s, state fish and game department crews demolished the Falk dam in order to facilitate passage of spawning salmon held in check by the dam's collapsed fish ladder. Today, school groups make field trips to Falk to see salmon traveling the Elk River during spawning season. The importance of water to fish, a seemingly unshakable cornerstone of nature, has nonetheless faced frequent challenges in both policy and practice. The relationship between the Elk River, the town of Falk, and the siting of the Elk River Mill was no different. Noah Falk's dam disrupted stream flow patterns and generations of indigenous fish, an occurrence common to natural resource industries of the time. (Courtesy of the Palmquist Collection, Humboldt State University Library.)

Policies requiring fishways to promote fish passage through artificially obstructed river channels date back to the 1700s. In the 1800s, some of the first studies regarding salmon populations and regional fishing practices were developed in response to the rise of commercial fishing and the effect of dams and habitat destruction. The first laws in California regulating fish and game date from 1851, and in 1872, the California State Legislature passed an act requiring fishways or hatcheries. "Fish commissioners" were appointed in the early 1870s, but initially received no salary for their work, which included inspecting dam infrastructure and conditions, including the installation and maintenance of fish ladders. In 1898, a Humboldt County newspaper reported the installation of a fish ladder around the dam at Falk's mill on the Elk River, presumably in response to this cycle of conservation policies. Falk's dam stood for nearly 60 years as a logistical boundary between downed logs awaiting the mill and spawning salmon looking for a home and as a representative feature of hard work and ingenuity in Humboldt County forestry practices. (Courtesy of Kristi Wrigley.)

These old-growth redwood logs were unloaded from the Falk No. 1, also known as the Gypsy, into the millpond. Once the trees were cut, they were transported by train to the Elk River Mill pond just upstream from the Falk mill. Logging operations at this time, in 1890, were about a mile east upstream from the mill in the South Fork Elk River watershed. (Courtesy of Jon Humboldt Gates.)

Prior to the completion of the Northwestern Pacific Railroad in 1914, milled lumber from the Humboldt Bay region was transported via steamships to the far reaches of the country, particularly San Francisco after the 1906 earthquake. The Bucksport & Elk River Railroad Company would take Elk River's lumber to the town of Bucksport, southwest of Humboldt Bay, at the present-day location of Bayshore Mall. (Courtesy of the Palmquist Collection, Humboldt State University Library.)

By 1910, the Elk River Mill and Lumber Company was producing 40,000 board feet per day and employed over 100 people. The mill had been in operation for 20 years and Noah Falk remained president of the company, but turned over daily operations to Winfield Wrigley as manager. The house located in the background was the home of Charles Falk, who had taken over as the superintendent in 1900. (Courtesy of HCC Photos, Humboldt State University Library.)

The town of Falk had a small general store, located near the mill, for which goods were ordered and transported from Eureka. For years, the company raised its own cattle and pigs for slaughter, but it was proved easier to order a weekly supply from local suppliers. Matt Carter, Falk General Store manager in early 1920s, was also responsible for ordering supplies for both Falk cookhouses serving approximately 200 men and the inventory of the Falk General Store. (Courtesy of Jon Humboldt Gates.)

Logging in Humboldt County began as early as 1850. As the town of Eureka transitioned, so did redwood logging methods. Early on, cutting down these giants was a drawn-out process taking two loggers a week with a single-sided ax. The axes were improved with double-cutting edges and longer handles. The crosscut saw was lengthened and used after the face cut. The back cut was made using a long crosscut saw with one faller on either side pulling through the cut taking turns. This work would take them several days at 10 hours a day. When the back cut was completed and the tree was about to fall, the loggers would jump off the springboards and watch the giant fall. The most infamous faller at Falk was Herb Christie, who is pictured on the right. (Right, courtesy of Larry Hill; below, courtesy of Jon Humboldt Gates.)

It is 1907, and this is "downtown" Falk, California. The entrance to Falk was along this plank road; on the right is the general store, and in the distance, one can see the steam coming from the Falk No. 1 train and the Elk River Mill. The house behind the mill belongs to Charles Falk, Noah's firstborn son, who would live in that home for the next 25 years. The town is situated in a narrow canyon, and placement of the buildings is strategic due to lack of space. The bachelor

cabins were next to the cookhouse, and the mill workers found it an easy commute to an early morning breakfast at 5:00 a.m. and work at 6:00 a.m. Today, this location is along the Elk River Trail in the Headwaters Forest Reserve, one mile from the trailhead. (Courtesy of Peter E. Palmquist Photographs, Yale Collection of Western Americana, Beinecke Rare Book and Manuscript Library.)

Christies Camp symbolizes the ownership Herb Christie had over both the woods and his crew as chopping boss. He was a skilled chopping boss who was able to estimate the exact location of where he wanted the trees to fall. His estimates were critical because smaller trees and bushes were placed on the ground acting as a cushion or bed once the tree fell. Without these cushions in place, the giants would splinter from their sheer weight and be useless lumber. In 1921, Christie was one of the highest-paid employees, making $120 a month with room and board. Walt Whitman described the early logging practices in his poem titled *Song of the Redwood Tree* from 1900: "The choppers heard not—the camp shanties echoed not; / The quick-ear'd teamsters, and chain and jack-screw men, heard not, / As the wood-spirits came from their haunts of a thousand years, to join the refrain; / But in my soul I plainly heard. / Murmuring out of its myriad leaves, / Down from its lofty top, rising two hundred feet high, / Out of its stalwart trunk and limbs—out of its foot-thick bark, / That chant of the seasons and time—chant, not of the past only, but the future." (Courtesy of Larry Hill.)

The woods crew was divided up into two groups—"the choppers" and the "machine operators." As logging continued deep into the Elk River watershed, the men took a 6:00 a.m. train ride up into the canyon and started their 12-hour day. Herb Christie (seated), from New Brunswick, Canada, started working for Elk River Lumber just after the mill opened in 1884. Forest depletion on the east coast of Canada sent many immigrants to work in the logging industry of the Pacific Northwest in the late 1880s. The logging companies welcomed these immigrants because they were experts in the field of harvesting trees. Also during this time in 1894, the United States entered an agreement with Canada to bring American ports of entry into compliance with inspections and proper immigration documentation. Prior to this time, many immigrants from Europe traveled to Canada and immigrated to the United States without being inspected or vetted through proper immigration laws. (Courtesy of Jon Humboldt Gates.)

In 1901, Humboldt County's mills manufactured and shipped 218,380,060 board feet of redwood, with the annual capacity of Noah Falk's Elk River Mill and Lumber Company being estimated at 15 million board feet, or about seven percent of the total for that year. Redwood products from Humboldt County included milled wood, shingles, and more. These goods were shipped to ports in San Francisco, Santa Barbara, Seattle, and beyond, including ports in Australia, Hawaii, and Japan. Humboldt County redwood products bound for export to distant countries faced stiff competition upon their arrival. Competition in these foreign markets came in the form of both native timber and timber obtained from other exporting countries. For example, in Japan in the 1920s, redwood products had to compete in an already established market that included Japanese

red cedar, Japanese white cedar, and several species of indigenous hardwoods, as well as logs and milled lumber from Russia, Borneo, and the Philippines. Moreover, timber exporters from the Pacific Northwest and British Columbia routinely shipped supplies of western red cedar, Port Orford cedar, Douglas fir, and hemlock to the industrial centers of Japan. Just as Noah Falk and his Humboldt County competitors had to find new ways to more efficiently get the redwoods from the forests to the docks, redwood marketers in distant countries had to find new ways to establish inroads into already thriving overseas markets. (Courtesy of Peter E. Palmquist Photographs, Yale Collection of Western Americana, Beinecke Rare Book and Manuscript Library.)

To say logging in the early 21st century was hard would be an understatement. Everything that the logger used was operated by hand. Bull and steam donkeys replaced oxen whose sole job was to get the logs out of the woods. The donkeys (pictured) were dragged out to a location inch by inch on skids. Once in place, the cables would pull the donkey around from log to log. The boilers in the donkeys ran by steam and water was a necessity. By the 1920s, much of the logging was done remotely and up steep ravines. Getting water to the donkeys proved difficult but not impossible. Water jacks brought water from nearby creeks and streams on horseback in large bags. One of the most difficult jobs was dragging the heavy cables back to the next log. They were weighty, stiff, and made of steel wire. It took many employees to complete a logging operation, something that would soon dwindle as automation grew. (Both, courtesy of Larry Hill.)

In the early days of logging, significant decisions were made just to get the logs transported to the mill. Once the railroad was erected, jack landings were constructed, a tree inventory was made, roads built, and the operation was underway. Bull donkeys (pictured) were the driving force in transporting the logs to various landings and were staged with a high lead system. The high lead had several cables attached to a spar tree (a center post) and was run through various blocks, to a cold deck. The whistle punk would blow a set of whistles alerting the bull donkey operator to go ahead and pull, go slow, stop or haul back. Paul Mazzucchi's job was to rig the cables around the logs and "run like hell" to a safe place away from the taut cables. (Both, courtesy of Jon Humboldt Gates.)

Transporting logs from the outlying woods without tractors and automation required ingenuity and multiple workers to operate log chutes, splash dams, and the railroad. In the log chute, there were several jobs, particularly the water jacks and the water slingers. The jacks had to pack water in two whiskey barrels per horse to supply water for the chute and the steam donkeys. In the summertime, it was challenging because, in just a few minutes, the heat evaporated the water, and the slinger was constantly throwing water on the chutes. Predictably, the water slinger was also in grave danger if the log gained too much speed, and the slinger had nowhere to run. Once the logs were down the chute and onto a landing, the donkeys would pull them up on the railroad cars to be transported to the mill for cutting. (Above, courtesy of Larry Hill; below, courtesy of Jon Humboldt Gates.)

The main jack landing where Noah Falk is standing is made out of horizontal logs, buried and anchored in the ground. This was an ideal surface for moving giant redwood logs on to the Falk Gypsy locomotive for transport. The steam donkeys rolled the logs onto the five cars on the train and transported them to the mill or out to Bucksport. Once the logs were brought into the mill, the lead sawyer would determine the best cuts and the mill workers would run them through the bandsaw. In 1921, the Elk River Mill employed 40 workers in positions such as head slab sawman, log hauler, carriage man, bolt cutter, and engineer. By 1928, redwood was being sold around the world to places like Japan as a sought-after exotic wood that was resistant to rot and perfect for outdoor construction. (Both, courtesy of Jon Humboldt Gates.)

Front to back, Albert McManus, Ed Callihan, and Harry Fryer pose on a redwood log bound for the saws in the Elk River Mill. Callihan was the lead mill sawyer for decades and oversaw how and when to cut. Albert was a steam donkey engineer and the son of Jim McManus, whose job title was "pond man." Jim's crew unloaded the logs into the pond and stacked them on the cold deck adjacent to the mill. On a day like any other, Jim was jacking up a log on the cold deck putting wedges underneath so it could be transported by the steam donkey. Albert, his son, was operating the steam donkey, and Jim signaled for him to operate the winch. The wedges underneath the log snapped and rolled over on Jim, crushing his legs. Immediately, they placed him on a flatcar headed for Bucksport. Emergency bells rang throughout town as residents gathered around the tracks. Jim saw his daughter Leila as he left town and waved to her but he never saw his daughter again. (Courtesy of the Callihan family.)

The manufacturing and distribution of redwood to distant markets all began with cutting the logs and transporting them to the mill. Here, the Falk No. 1 Gypsy engine is being used to off-load logs for temporary storage in the Elk River Mill pond. Redwood logs were not kept in the pond indefinitely but moved to a cold deck further down the transportation chain until they were needed for the day's milling operation. Logs were stored in a pond to soak off dirt, small rocks, and other debris that had accumulated on the bark of the logs as they made their trip from stump to storage. This type of debris could wreak havoc with the mill saw blades. Mills such as Elk River employed peelers to debark trees prior to milling. Jim McManus, the Elk River Mill and Lumber Company pond man, would maneuver the logs from the delivery point to the mill. He had caulk boots and a long pole or peavey for balance. Despite earning only $70 a month in 1921 wages, McManus and pond men throughout the Pacific Northwest provided this critical yet dangerous service. (Courtesy of the Clarke Historical Museum.)

Room and board at Falk averaged $5 a month. Considering the workers made $2.50 a day and worked six days a week, this was a fair price for three meals a day and a cot. Many of the single men would share a cabin and play cards in the evening. Paul Mazzucchi remembers working for the Hammond Company and said after 10 hours of work, he would come back to his cabin and there would be clean white sheets on his bed with a fire going in the woodstove. He said it was worth every penny to come back to a dry, warm place with a stomach full of cookhouse food. (Left, courtesy of Jon Humboldt Gates; below, courtesy of the Boyle Collection, Humboldt State University Library.)

A wage and occupation ledger from November 1921 shows that the Elk River Mill and Lumber Company employed 119 workers. Wages in 1921 ranged from $40 per month to $250 per month. Along with their wages, 112 of the 119 employees received full board, reflecting a key aspect of the company town model of that time. By March 1923, the wage and occupation ledger revealed that employment at the Elk River Mill and Lumber Company exceeded 150. Wages were notably higher in 1923 due to the economic boom of the Roaring Twenties. The lowest-paid staff members now earned $55 per month plus board and a discernible shift toward higher wages dominated both skilled and unskilled positions. From 1921 to 1923, wages for Elk River Mill and Lumber Company log peelers increased from $90 per month to $100 per month. The camp flunkey was an even bigger beneficiary, jumping from $40 a month in 1921 to $70 a month in 1923. (Right, courtesy of BLM; below, courtesy of Larry Hill.)

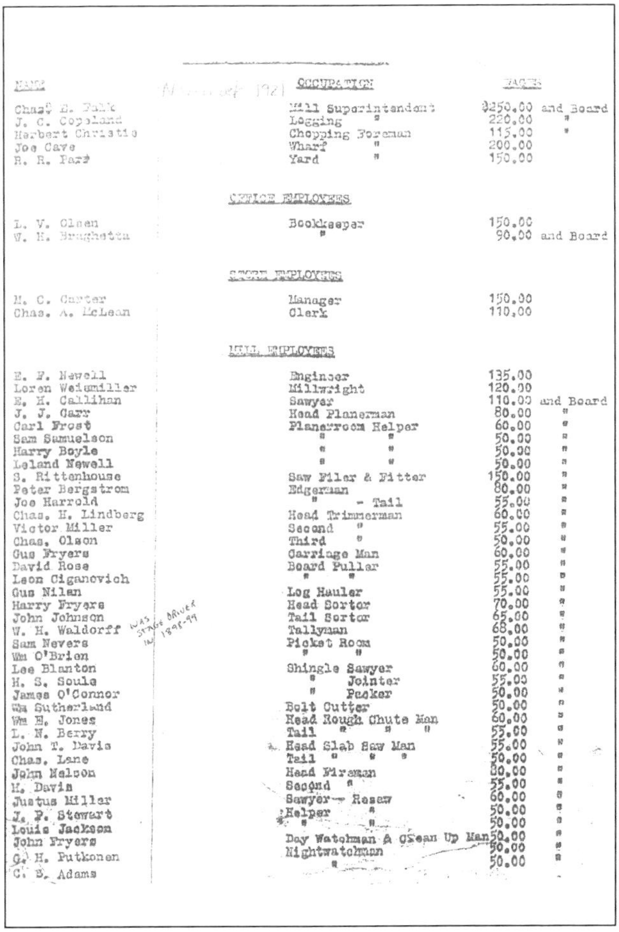

November 1921

NAME	OCCUPATION	WAGES	
Chas. E. Falk	Mill Superintendent	$250.00	and Board
J. C. Copeland	Logging "	220.00	"
Herbert Christie	Chopping Foreman	115.00	"
Joe Cave	Wharf "	200.00	
R. R. Parr	Yard "	150.00	
	OFFICE EMPLOYEES		
L. V. Olsen	Bookkeeper	150.00	
W. H. Brughetta	"	90.00	and Board
	STORE EMPLOYEES		
H. C. Carter	Manager	150.00	
Chas. A. McLean	Clerk	110.00	
	MILL EMPLOYEES		
E. F. Newell	Engineer	135.00	
Loren Weidmiller	Millwright	120.00	
E. H. Callihan	Sawyer	110.00	and Board
J. J. Carr	Head Planerman	80.00	"
Carl Frost	Planerroom Helper	60.00	"
Sam Samuelson	" "	50.00	"
Harry Boyle	" "	50.00	"
Leland Newell	" "	50.00	"
S. Rittenhouse	Saw Filer & Fitter	150.00	"
Peter Bergstrom	Edgerman	80.00	"
Joe Harrold	" - Tail	55.00	"
Chas. H. Lindberg	Head Trimmerman	60.00	"
Victor Miller	Second "	55.00	"
Chas. Olson	Third "	50.00	"
Gus Fryers	Carriage Man	60.00	"
David Rose	Board Puller	55.00	"
Leon Ciganovich	" "	55.00	"
Gus Nilan	Log Hauler	55.00	"
Harry Fryers	Head Sortor	70.00	"
John Johnson	Tail Sortor	65.00	"
W. H. Waldorff (Was stage driver in 1898-99)	Tallyman	68.00	"
Sam Nevers	Picket Room	50.00	"
Wm O'Brien	" "	50.00	"
Lee Blanton	Shingle Sawyer	60.00	"
H. S. Soule	" Jointer	55.00	"
James O'Connor	" Packer	50.00	"
Wm Sutherland	Bolt Cutter	50.00	"
Wm E. Jones	Head Rough Chute Man	60.00	"
L. N. Berry	Tail " " "	55.00	"
John T. Davis	Head Slab Saw Man	55.00	"
Chas. Lane	Tail " " "	50.00	"
John Nelson	Head Fireman	80.00	"
H. Davis	Second "	55.00	"
Justus Miller	Sawyer - Resaw	60.00	"
J. P. Stewart	Helper "	50.00	"
Louis Jackson	" "	50.00	"
John Fryers	Day Watchman & Clean Up Man	50.00	"
G. H. Putkonen	Nightwatchman	50.00	"
C. B. Adams	"	50.00	"

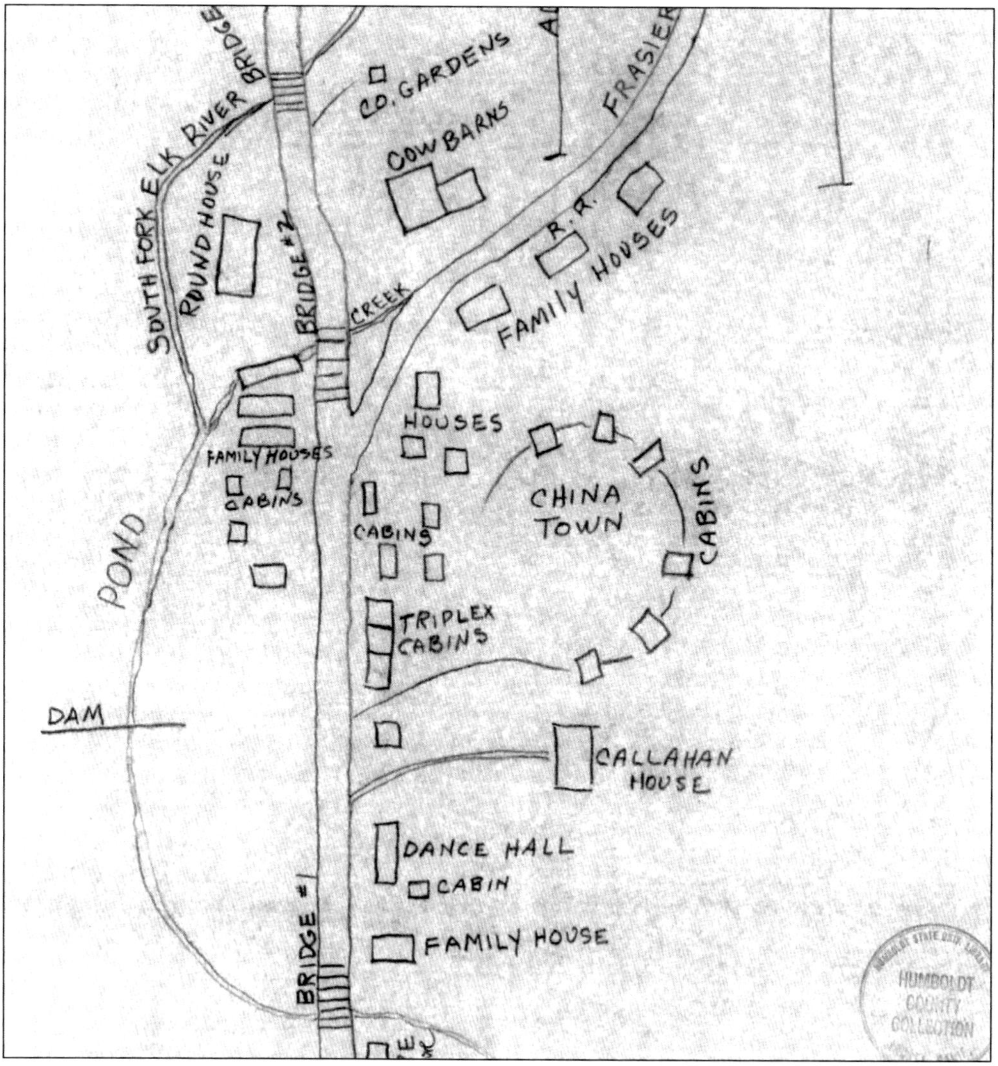

Although there were many immigrants that came to work for Noah Falk, they were all of European origin. On a historic map of the Falk townsite, there is a section of houses labeled "China town." These groups of houses were bachelor cabins belonging to loggers and mill workers who labeled it as an inside joke. In 1885, just after Elk River Mill was constructed, the infamous Chinese expulsion of Eureka's entire Chinese population, which numbered about 300 men and 20 women, took place. Businessmen and residents rounded up the Chinese and brought them down to two steamships in Humboldt Bay after two rival Chinese gang members accidentally shot a Eureka city councilman, Dave Kendall. A year later, in 1886, Eureka papers announced the coming anniversary of the tragic death of Kendall and pleaded with citizens to renew their pledge on keeping the Chinese out. Subsequently, a meeting was held in Arcata, and Falk and Isaac Minor spoke against the Chinese and said they should leave immediately. In the early 1890s, the front page of a local history and business directory declared in boldface type that Humboldt was "the only county in the state containing no Chinamen." (Courtesy of the Humboldt Room, Humboldt State University Library.)

These 1910 Falk woodsmen are proudly posing in front of a large redwood split log. It could take three or more days for a two-man team to fell a 12-foot-diameter tree. Larger logs were often split into manageable sizes, sometimes using explosives. When large crosscut whipsaws were developed, a crew might cut one or two large trees a day. (Courtesy of Jon Humboldt Gates.)

Humboldt County mills and mill workers handled redwood logs as large as 20 feet in diameter, logs so large they sometimes had to be split with dynamite before being further processed. Saw blades and milling technology evolved rapidly to handle these large-diameter logs, with circular saws replacing sash and muley saws in the 1860s and double circular saws seeing widespread use in larger mills by the 1870s. By 1921, mill employees received between $50 and $150 plus board per month. (Courtesy of Jon Humboldt Gates.)

Leading-edge technology may not be the phrase most commonly associated with the oxteams and sawdust-filled air of 19th-century logging, but it applies directly to the operations of Noah Falk and the Elk River Sawmill. At the time, tools to mill trees often came from Europe or the eastern states, where average log diameters were only a few feet. In fact, William Carson, another Humboldt County logging pioneer, skipped over the large-diameter redwoods when delivering his first load of milled lumber, cutting only the smaller trees that his tools could handle. But Noah Falk turned to the bandsaw to run his mills and insisted on steel components for durability. Trade journals of the day credited Falk as a forerunner in the use of bandsaw technology and acknowledged that his Jolly Giant and Elk River Mills as the first to achieve commercial success with bandsaw logging blades. Competitors took note of Falk's advances and adopted the bandsaw in their own mills. (Courtesy of HCC Photos, Humboldt State University Library.)

The water jacks pictured above provided the water for the Dolbeer steam donkey that replaced the oxen pictured in the below photograph. They would either carry the water in large sacks (pictured) or in barrels on top of their horses and mules. In 1870, the 14-foot-diameter old-growth tree seen below was felled in Arcata near Jolly Giant Creek by loggers who worked for both Isaac Minor and Noah Falk. The two men were business partners and had a competition going on who would make a bigger name for himself in the community of Arcata. Minor wanted to build a hotel to be remembered by; however, someone suggested he open a movie theater. The Minor Theater was opened December 3, 1914, and is the oldest building in the United States designed for feature films. (Courtesy of HCC Photos, Humboldt State University Library.)

In 1907, during the time when this photograph of the Falk woods crew was taken, Noah Falk was known by his employees as a just and respectable man who gave his employees fair wages, board, and ample work. During this period of American history, mass immigration from southern and northern Europe dominated company towns. Living conditions varied across the country, and the owner was also the food supplier, the post office, the bank, and the source of entertainment. This was a breeding ground for exploitation and paternalism that gave the proprietors the upper hand, trapping employees in a circle of debt, which Merle Travis put so eloquently in a song titled "Sixteen Tons": "I owe my soul to the company store." When local labor strikes broke out in Humboldt County in 1935, Elk River Lumber Company was not one of the offending paternalistic companies. (Both, courtesy of Larry Hill.)

The neighbors downstream from Falk faced a variety of issues, including damages from floods and poor water quality. In February 1921, large logs that were stored on the cold deck were thrown in the water behind the millpond dam. Meanwhile, a significant amount of water came down the Elk River, causing the water level to rise, and the weight of the logs washed the dam out. Most of the logs came down the river for at least a mile and a half and ended up at Paul Mazzucchi's parents' property. Mill superintendent Winfield Wrigley said to the Mazzucchi family, "They're company logs, I want them back." They refused to give them back, so Noah Falk met with Maddalena Mazzucchi, and Noah told her the company suffered quite a loss and that it was "an act of God." Maddalena asked, "What about the little guys down the river from you? The logs are mine and if you want them you buy them off me, you can pay for the damages too." They were able to sell the logs and build the house that sits on Elk River Road today. (Above, courtesy of HCC Photos, Humboldt State University Library; right, courtesy of Jon Humboldt Gates.)

From 1889 to 1894, just west of what is now the Headwaters Elk River parking lot, was the town of Scribner. Leila Scribner opened a post office variety store after her husband, Benton, was killed after a serious logging accident. Benton was a hook tender, who attached cables to the down logs and signaled the steam donkey to pull. This was the most dangerous jobs in the woods. When the cables snapped at high pressure, they would fly aimlessly and could go right through the nearest person. Benton placed the hooks on the log, and the hooks came loose and hit him in the head. He died the next day, leaving his wife, Leila, a widow with two small children. There was no life insurance, no compensation from the company, and no money prospect for Leila. Being a resourceful person, she saw a need for a post office and a stationery store near the newly developed town of Falk. Her post office closed in 1894, and she remarried in 1898, moving to Eureka proper. A new post office was erected at the Falk townsite (pictured). (Above, courtesy of Jon Humboldt Gates; below, courtesy of Larry Hill.)

When Noah Falk left the Dolly Varden Mill in Arcata in 1886 and set up shop in Elk River, he brought with him the Gypsy locomotive, also known as the Falk No. 1. This was a little engine that could, hauling old-growth logs that far outweighed the train on several cars downstream to the millpond. Noah eventually purchased 3,000 acres upstream and had to build a rail line to it. The steep grade was too much for the Gypsy and a larger more powerful Baldwin locomotive was purchased. The Gypsy steam locomotive was near and dear to the hearts of the Falk residents, and the owners kept it around to work the millpond and run errands from Falk to Bucksport. It was retired in 1927 and ran briefly on the streetcar tracks of Eureka for General Grant Days in 1937. The engine was restored by North Counties Logging Interpretive Association, now the Timber Heritage Association, in 1977. It was taken to the world's fair in Vancouver, British Columbia, in 1986 and presently sits at Fort Humboldt. (Above, courtesy of the Clarke Historical Museum; below, courtesy of Jon Humboldt Gates.)

Noah Falk came to California in 1854 at the age of 17 with little money in his pocket. By the time he was 47, he was one the wealthiest men in Humboldt County, owning and operating several different mills including Elk River Mill. His brothers Elijah and Jonas from Ohio joined Noah in 1878 and were a large part of his success. Elijah had a sawmill business in Ohio and had learned how to be a millwright before he ever came to California. In 1884, Noah gave Elijah the job of designing the Elk River Mill and seven other mills. Elijah revolutionized the shingle industry in the United States by designing and building the first shingle drying plant in Bucksport. After his retirement in 1915, Elijah became the first Socialist mayor in Eureka. Noah Falk passed away at the age of 91 in 1928 after complications from being hit by an automobile while crossing the street in Arcata. (Both, courtesy of Jon Humboldt Gates.)

This 1903 photograph shows Noah Falk's view from Twelfth and H Streets to the southwest and overlooking the Arcata bottoms. Notice the ornate fence in the immediate foreground, the Methodist church, and the large vegetable garden across H Street. During this period, Union (Arcata's original name) was an affluent community with many lumber barons such as Noah Falk and Isaac Minor. However, with limited resources due to Arcata's remote location, many of the residents grew their own vegetables and canned produce for winter months. Dry goods and groceries were purchased at Brizard's, present-day Jacoby Storehouse. (Courtesy of the Palmquist Collection, Humboldt State University Library.)

Built by contractor Theodore Dean in 1887, this ornate Victorian stood on the corner of Thirteenth and H Streets, Uniontown, Arcata (present-day Wildberries Market). By 1903, this was home to Noah and Nancy Falk. Noah was in his late sixties and had stepped back from day-to-day operations at Falk. As an influential Arcata businessman, he partnered with lumber baron Isaac Minor. Falk was one of the leading citizens of Humboldt County and called the pioneer of the "Redwood Realm." Note the elaborate fencing on top of the roof. (Courtesy of the Palmquist Collection, Humboldt State University Library.)

Having been a resident of Humboldt County for 60 years, Noah Falk became a prosperous lumber businessman employing hundreds of employees in several mills. His interests were large and diversified. By 1920, he had been a banker, stockholder and director in four Humboldt County banks—Bank of Arcata, Arcata Savings Bank, Bank of Eureka, and Humboldt County Savings Bank. He was also the president of the Hotel Arcata as well as the Union Water Company in Arcata. He died in March 1928 at the age of 92. (Courtesy of HCC Photos, Humboldt State University Library.)

Noah and Nancy Falk (seated in front in the car on the right) are pictured in their 1903 motorcar driving the streets of Arcata at a time where very few people had an automobile. Noah Falk's influence and generosity was paramount to the development of early Arcata. As a philanthropist, he donated firefighting equipment to the local fire department and, in 1908, a plot of land to the Arcata's Women's Club and ultimately was responsible for the donation of land to the City of Arcata in creating Redwood Community Forest. (Courtesy of the Palmquist Collection, Humboldt State University Library.)

When Charles Falk retired in 1930, Winfield Wrigley took over the position of Elk River Mill lumber superintendent. Shortly after this time, the company had just a handful of employees and was basically shut down for six years. The company was in desperate need to make the necessary upgrades to compete with the larger local logging companies. The lumber that was coming out of Falk was cut on steam-powered bandsaws, and the logs were still being moved by steam donkeys instead of tractors. In 1936, Winfield secured a $1 million government loan to upgrade the donkeys to crawlers that sped up logging operations. He lived in what was once Noah's, and then Charles's, house overlooking the mill (pictured). Owner John Reed, who lived in San Francisco, called Winfield in 1937 and officially shut down the mill due to high operating costs and low volume. The operation was shut down for good and the company went into bankruptcy. (Courtesy of Jon Humboldt Gates.)

By 1937, when Elk River Mill closed, the loggers were still using crosscut saws and axes. It was not until World War II that demand for wood products exceeded obtainability. When the war started, chain saws still were not common in the forests of the Pacific Northwest. However, that changed rapidly when the call was heard loud and clear that wood products were needed to build barracks, hangars, ships, bridges, and crates. A propaganda poster of the time captured the get-out-the-cut urgency: "Lumber production is falling behind our war needs. To save American soldiers' lives, we must provide the lumber our armed forces need—now!" The chain saw rapidly changed the amount of time needed to cut down trees; one faller could do the work of 15 fallers who had axes and crosscut saws. Historic resource extractions changed with demand and technological advancements. (Courtesy of Wayne Miller.)

After World War II, the demand for Douglas fir and redwood lumber increased due to the housing boom. After 1950, redwood trees fell triple the rate than they had previously. The advent of new technologies such as the chain saw (pictured) accelerated logging. Before the war, a rudimentary chain saw required two people—one to hold the engine and one to hold the handle by the bar. Wayne Miller and Ted Williams (pictured) both worked for Hammond Lumber Company in the Elk River watershed using one of the biggest bars on a chain saw for 1953. By the end of the 1950s, only 10 percent of the original two-million-acre redwood range remained untouched. Fortunately, due to the efforts of conservationists as early as 1900, old-growth redwood trees from Big Sur to Brookings, Oregon, are protected and can be seen and treasured for generations to come. (Courtesy of Wayne Miller.)

"Clear Redwood Planks Without a Knot or Flaw" was certainly boosterism at its finest, yet this quote from the 1904 *Humboldt County Souvenir* was not at all far from the truth. Redwood products were of world-class quality, easy to work with, pleasing to the eye, and seemingly miraculous in resisting insects and weathering. For nations lacking a similar fiber resource, redwood answered the need for a building material that offered longevity and fire resistance, both highly desirable characteristics for the time period. Sawed lumber represented the largest share of the forest products recorded in the datasets found in this turn-of-the-century promotional piece. Shingles, shakes, railroad ties, pickets, posts, staves, doors, moldings, and sashes were also listed as high-value forest products, along with butter boxes and butter molds. The overall value of these products was listed as being nearly five million dollars for the production year. Despite a rather rough outward appearance, Falk woodsmen and milling crews were key participants in a sophisticated global economy. (Courtesy of Jon Humboldt Gates.)

Two

Falk Community

The community in Falk can best be described as a tightly woven fabric with diverse cultures, economic status, and a past that brought people together in a small river valley surrounded by giant trees that were yet to be logged. Bound by work and the chance for a better life in America, men came to Falk for the American dream by emigrating from places like Europe, Texas, and the far reaches of Canada. Whether they brought their families or traveled alone, they all had one thing in common, a desire to work hard, along with a willingness to work 10-hour days in a hazardous landscape of falling redwoods, tight cables, sharp saws, and exhaustion. The limited available space of the Elk River valley made for an intimate setting with an atmosphere of an unspoken caste system, isolation, self-reliance, many hands, and a lack of modern conveniences. Children growing up together depended on each other for entertainment, tutoring, gossip, and even a future mate. The social network from the earliest age consisted of neighboring families who picnicked together, went to church together, traded fruits and vegetables and recipes, and helped one another. Whether good or bad, community life was hinged on the survival and happiness of one another. Modern conveniences such as the automobile, electricity, and television forever changed the composition that the community once was. Fortunately, the public land that is now Headwaters Forest has a new community of stewards who can leave those modern conveniences behind and step back in time for a little while, finding respite in the silence of the forest.

Company towns were convenient places for new immigrants in the early 1900s. Much of the lumber industry workforce came from countries such as Ireland, Finland, Norway, and Sweden, and the Canadian province of Nova Scotia. For immigrants who did not speak English, chances are someone in town spoke the same language and ate the same food. Falk, as many company towns, was an easy place to navigate. Food, water, shelter, and community was just an arm's reach from the place of employment. After work, many of the loggers and mill workers came together for meals, card games, knife-throwing contests, and Sunday laundry day. (Both, courtesy of Jon Humboldt Gates.)

In the early 1900s, substantial migration both foreign and domestic was an important part of company town populations. The Barnes family came overland from the Sacramento region of Woodland. Charles "Pete" and Loleta "Lete" (pictured far left) raised their two sons Chester "Ted" (left center) and Marshall (center). They had a vegetable garden, several cows, and chickens. Lete would sell the surplus milk from her cows to neighbors for extra money. She would eventually live in the town well after it closed and become the caretaker until she passed in 1975. Ted was an enterprising young boy who would sell mixed drinks at the Falk dance hall. By doing so, he made more than a week's labor for a foreman in the woods. Ted graduated in 1924 from Eureka High School. Many of the Falk boys left school by eighth grade to go to work. Ted eventually worked for Carson Lumber Company until he was injured twice in one year and went on to be a truck driver until he retired. (Courtesy of Jon Humboldt Gates.)

Jack Miller made it out west from Texas to Arizona in 1899. He worked as a farrier and was taught blacksmithing skills for a couple of years until he made it out to San Francisco. Apparently, he had enough money to buy buttoned leather shoes and a one-way ticket to Eureka. He was so proud of the shoes, until he stepped off the curb in Eureka and sank in the mud. Miller was offered a position as the Falk blacksmith in 1904. He married Ruby Furlong and had two children, Evande and Wayne. The blacksmith had one of the most important jobs in the mill. He made and manufactured all the moving parts necessary for mill operation, shoed horses, and made all the parts necessary for logging operations. Blacksmiths often worked long hours, even into the night. Jack made $120 a month plus room and board in 1921. (Above, courtesy of Wayne Miller; below, courtesy of Jon Humboldt Gates.)

Just because one grew up in Falk, lived an entire life next to the mill, and had parents who worked in the mill for decades, no compensation was given due to an injury. Evan "Ev" Rushing II, born in 1897, grew up in Falk and worked in the blacksmith shop for a short time, and then he and his brother Lloyd started working in the woods. Ev's job was as a zoogler. He was responsible for the rigging cables around the logs that were pulled around by the donkeys. On a routine pull, the cables got snagged on a bridge, and Ev jumped down and cut the cable with an ax. The cable whipped around and snagged his ankle, severely injuring him. He was not able to walk or work for almost a year. (Courtesy of Jon Humboldt Gates.)

Ev Rushing had no choice but to be treated for his injury at the local hospital. He did not have the resources to pay his medical bills under what is today workman's compensation. Under the Workman's Compensation Act of 1913, the Boynton Act carried forward the provisions of the Roseberry Act, imposing liability for compensation "without regard to negligence." An employee injured by his or her own willful misconduct was still denied compensation. As under the Roseberry Act, an employee was permitted to choose a civil damage action rather than statutory compensation. Rushing did not sue Elk River Lumber Company, but the hospital did, for $1,000. A year later, Ev was invited to come back and work for the company. Later that summer, John Reed, the owner, came up from San Francisco and saw Ev engineering the train. He ordered Winfield Wrigley to fire Ev as a liability to the company. Later, Jim Copeland related that he had to let one of the best woodsmen go. During this time in history, unfair labor practices were common and getting injured was often met with getting fired. (Courtesy of Jon Humboldt Gates.)

The Aho family were Finnish immigrants who came to Humboldt County in the early 20th century. Emil and Elizabeth were neighbors in Finland but did not know it until they both met in Eureka and married. Alli Aho (pictured with brother Emil) was born in March, a month before the San Francisco earthquake of 1906. When Alli was born, her father Emil worked in the Elk River Mill and Lumber Company as a woodsman felling trees. A year later, her family moved into the town of Falk because her father took a better-paying job as a night watchman in the mill. He worked six days a week, and the Aho family lived comfortably on his wages. The family lived in a simple wooden house with a kitchen, two bedrooms, and a living and dining room. It was situated on a small half-acre lot with room for a chicken coop, a vegetable garden, and yard. Inside, kerosene lamps lighted the house, and a wood-burning stove heated it. One Christmas morning, Alli and her brother woke up to find a decorated tree in the living room with colorful rosettes their mother had made out of tissue paper. This was a memory that Alli never forgot, and she always wondered how her mother managed to make that Christmas so bright. (Courtesy of Jon Humboldt Gates.)

The population and diversity of Falk immigrants were widespread during its 50 years. Residents heard many different languages and dialects throughout the workday. The pull to Falk was the plentiful work in both the woods and the lumber mill. The push from European countries was the lack of work and opportunity for growth. Archaeologists excavating the housing sites discovered that the cultures reveal themselves in the garbage that was left behind. Franz Goessi traveled to America from Switzerland on a ship where he met his wife, Bertha, in 1892. They were two of 12 million immigrants who passed through Ellis Island between 1892 and 1954. Franz found work as the Elk River Mill's fireman, whose job it was to keep the boilers fed with sawdust. They went on to have eight children and eventually moved to Elk River in 1899 after they purchased six acres for $450. In 1908, Bertha died after giving birth to her eighth child. Frank became a widower and a single father who made money farming and selling strawberries and cream to passersby on Sunday as well as from his work at the Elk River Mill. (Courtesy of Sheldon "Zeke" Sicotte.)

Finnish interest in America began to develop in the mid-19th century when the news of the discovery of gold in California trickled to Finland. Finnish seamen left their ships when they docked in American harbors and never returned. Many of the immigrants were drawn to the logging industry in Washington, Oregon, and California. Martin Fleming emigrated from Finland in 1880 and was drawn to the lumber industry due to his previous experience. Martin met his wife, Hilma Pyykola (also Finnish), in Eureka, and they went on to have six children. Martin and his family moved to Falk around 1910. Hilma died of milk fever shortly after her youngest son, Milton, was born. Martin had six young children, including Helen (pictured) that needed to be cared for while he worked and relied on his community and other families during this difficult time. He worked as a mill worker, steam donkey engineer, and a saw filer. He continued to live at Falk well after the mill closed into the 1940s. (Both, courtesy of the Callihan family.)

Company town cooks had enormous power in both their management of their crew and their leverage in providing great food. Maggie McNeill (center) worked three miles upriver from Falk at a cookhouse called "Maggie's Camp." Her main goal in life was to serve the best possible food to the loggers who entered her cookhouse. She did not like men complaining about her food and would bribe children to spy on people who did. The kids would report back to her, and she would not let the offending complainer in her cookhouse for the day. She had a reputation for a fiery temper and would send food products not to her standard back to general store manager Matt Carter. Her talent as a cook brought about a strike when she took a leave of absence, and the succeeding two cooks were marginal at best. McNeill is pictured on the deck of her cookhouse with her hardworking kitchen crew in 1910. She worked for several years in logging camps, married three times, had one son, and died at the age of 70 in 1952. Pictured from left to right are (first row) Ruby Barnes, Suzi Glass's daughter, and James Biord; (second row) Bernice Barnes, Maggie Biord-McNeill, and Suzi Glass. (Courtesy of Jon Humboldt Gates.)

The cookhouse, also known as Maggie's Camp, was built around 1922 and was approximately two miles upriver from the Elk River Mill. The loggers would take the train from Falk and head up the canyon around 6:00 a.m. for breakfast. The cooks started their workday at 3:00 a.m., prepping for a meal that would carry the logger's energy into lunch. Breakfast consisted of steak, eggs, potatoes, pancakes, mush, and coffee. Paul Mazzucchi remembers that after breakfast one person from each group of workers picked up a pre-filled bucket containing raw meat, potatoes, and other items. That worker would break off during the 10-hour work day, just before lunch, and literally prepare a meal suitable for a hungry lumberjack. Practical jokes were commonplace, including sneaking in before breakfast and putting rocks in the bottoms of the lunch buckets. (Courtesy of Jon Humboldt Gates.)

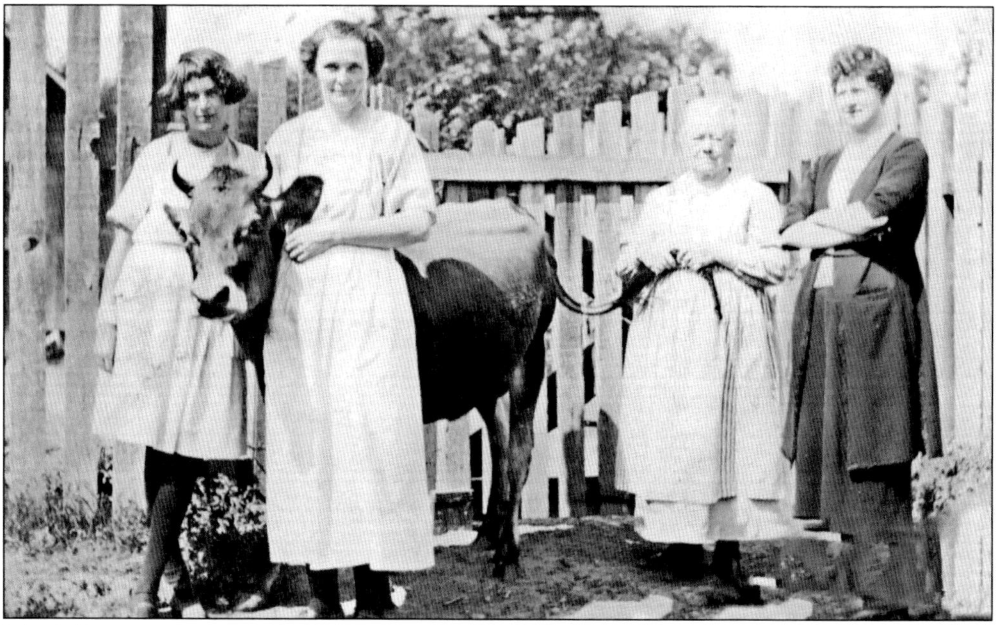

The Falk community was rich in layers with multiple generations sharing one household. Yearlong residents proved to be self-sufficient in that each family had a cow, chickens, pigs, and a vegetable garden. Chores were divided up between family members, and no one was ever too old or young to contribute to getting water, cutting wood, sewing, or cooking. During the Depression, many of the residents relied on the food that grew naturally around them, dandelions, salmon, and deer. Dorothy Fleckenstein recalled never wanting to eat salmon or deer again. The Elk River watershed was a fertile area to live off the land during times of necessity. The neighbors traded and shared fruit trees, ornamentals, and canning products. Today, along the Elk River Trail at Headwaters, there is evidence of this time period in the fruit tree blossoms in the springtime. (Above, courtesy of the Callihan family; below, courtesy of Jon Humboldt Gates.)

Women began working in logging camps in the Pacific Northwest as early as the 1900s. Often, they would travel long distances to work as a camp flunkey until they garnered enough experience to be a camp cook. Their job began before sunrise and lasted well into the evening and was packed full with jobs such as wood splitting, potato peeling, dishwashing, waitressing, and cooking. Single-woman status in company towns was short-lived. Women would sleep in their "flunky shacks" (off-limits to men), comprised of cots and a wood-burning heater, and converse well into the evening about what they would wear for that Saturday night dance and whom they wanted to dance with. It was at these dances that the single-men loggers, brake tenders, hook tenders, and donkey operators would ask for a dance with the ladies who caught their eyes that previous week in the dinner shed. Whether working in the cookhouses or raising a family, the women struck a balance of hospitality within the life of a company town. (Both, courtesy of the Callihan family.)

Many Falk houses were located at the south end of Elk River. To access their homes, residents parked in small garages located on what is now the Elk River Trail and crossed small footbridges (pictured). Carrying home building supplies across these bridges must have been daunting. Lottie Callihan is pictured on the porch of her house that was located directly across the mill. She and her husband Ed married in 1891. He was the lead sawyer at the Elk River Mill and had one of the most important jobs. He had to calculate how to cut and where to cut by looking at the grain of the log and estimating board feet. He worked in the mill for decades and suffered later in life with leg problems from excessive standing. Ed passed in 1934, and Lottie joined her daughter Zelda in Los Angeles. (Both, courtesy of the Callihan family.)

Dance halls and other forms of entertainment were instrumental in the development of company towns. The owners knew the value of celebration and recreation among the community members. Falk was famous for its dance hall. Bands like Cole's Californians would play at different dance halls throughout Humboldt County. This was a chance for the women to put on their best clothes and bring their best food dishes and for the men to unwind from a six-day workweek. Ruby Fryers taught the men how to dance, and, apparently, they walked all over her feet. She also played the piano, and Joe Cox played the banjo. When the automobile was more accessible to Eureka residents, they would come out to Falk and enjoy a midnight supper of fried chicken, mashed potatoes, and homemade pies, followed by dancing until sunrise. The men worked six days a week, and this was their only chance to stay up late and not have to worry about working the next day. (Right, courtesy of the Callihan family; below, courtesy of the Boyle Collection, Humboldt State University Library.)

Growing up in a company town was infinitely different than growing up in a typical working-class neighborhood. The concentrated small-town atmosphere lent itself to a closeness that was both comforting and confining. Paul Mazzucchi remarked that everyone knew each other's business, and privacy was in short supply. On the positive side, children growing up in company towns had someone to share their play, work, and pastimes similar to an extended family. Lewis Olsen recounted that the woods were their playground, where they could fish, swim, build forts, and play army with each other. The Falk children formed relationships that went on well after the mill closed and into their elderly years. Many of these stories are captured in Jon Humboldt Gates's book *Falk's Claim: The Life and Death of a Redwood Lumber Town*. (Both, courtesy of the Callihan family.)

Lewis Olsen (right) and Ray Haley sit on the back porch of a Craftsman-style house with Olsen's dog Spot. Apparently, Spot howled at the moon and killed garter snakes in the surrounding forest. Olsen and Haley were neighbors in 1925 when the last Falk house was built. It was also the first house to have plumbing and the last house to survive, collapsing in 2012. Lewis's father, Vernon, was the bookkeeper at Falk from 1913 to 1932. He married Ruby Barnes, who worked in the cookhouse with the infamous cook Maggie. Vernon worked many unofficial jobs in his life, including taking a town drunk who lost his leg to the hospital and serving as the unofficial bank of Falk. Many times, he lent workers their money before payday, and in all his years, only one man skipped town without paying back the $20 owed. Lewis fondly remembers his dad bringing him home a bag of peanut clusters after his shift in the store. (Right, courtesy of Lewis Olsen; below, courtesy of BLM.)

Company towns employed workers but they also came with their families. Children were in need of suitable schools and compulsory education was in effect. Jones Prairie School was located about a mile from Falk, and the teacher everyone remembered was Alice Wrigley. She was the kindergarten-through-sixth-grade teacher at Jones Prairie School for nearly 45 years. In 1910, when she began teaching, a second school next door was built with full attendance of 100 students. All six grades were taught in one room, and the older students often taught the younger grades. There was a small playground to the right of the school with five swings and a set of rings. Wrigley was a typical schoolmarm of the day who never married and devoted her life to education. Wayne Miller, the Falk blacksmith's son, never forgot the day that Wrigley came over to speak to his parents on a discipline issue. Instead of sending notes home, she would just arrive at the house to have a face-to-face conference with the parents. (Both, courtesy of Jon Humboldt Gates.)

In 1900, at the age of 10, Maude Frost was living with her widowed mother on Carson's property near Falk with her six brothers and sisters. Maude's father was killed when he was splitting shingle bolts and the froe (splitting tool) hit him in the neck. The eldest brothers Carl (14 years old) and Ralph Frost (16 years old) worked as laborers in the Elk River Mill to support the family. Maude Frost and Alice Wrigley both graduated with their high school diploma/teaching credential in 1909 from Eureka High School. Maude (below, back row, center) taught at Jones Prairie School from 1913 to 1915 and from 1920 to 1923. Paul Mazzucchi, a resident of Elk River, recounted a story about Maude. He said when he was 15, she told him that she had taught him everything she knew and he was old enough and strong enough to leave school and go to work. She got him his first job as a rigging operator, setting cables around the logs to be pulled by steam donkeys. He worked in one of the most dangerous logging positions for almost 20 years. (Both, courtesy of Jon Humboldt Gates.)

Seen after a Sunday service, the First Congregational Church sat on a hill behind Jones Prairie School; the cabin in the forefront was the parson's cottage. This church was the only available place of worship within miles of Falk. Otherwise, families would need to take a long buggy ride into Eureka or spend the day at home reading the Bible with family and friends. Herb Christie, Elk River's lead chopping boss, was also responsible for leading Sunday school classes in the church. Nearly 60 children attended and received Bible lessons and had a chance to sing with friends prior to attending the formal service with their parents. With the church in place, it gave Falk residents a more complete community—with a school, post office, dance hall, market, and now a place to worship on Sundays. (Courtesy of Jon Humboldt Gates.)

By 1912, there were two schoolhouses, one for the primary grades and the other for the middle school children. There were four teachers during this time: Alice Wrigley, Mary Weatherby, Myrtle Barnum, and Luella Van Horn. Judging by the attendance at both schools gives one a clue about the success of the community and mill operations. The more children, the greater the probability their parents were successfully employed and working at jobs that adequately provided for the family. In 1912, across the United States, children over the age of 12 worked alongside adults in factories and on farms, especially in the Elk River Valley. By 1938, the first federal child labor law was passed, allowing teens between the age of 14 and 16 to work limited hours and those older than 16 to work longer hours. By this time, Jones Prairie School had been open since 1887 with limited attendance. By 1964, Jones Prairie had one teacher, Mary Lengal, and closed its doors permanently. (Both, courtesy of the Clarke Historical Museum.)

Jones Prairie School, named after D.R. Jones, who first logged the Elk River Valley, was located one mile downriver. Adjacent to what is now the Headwaters Forest Reserve, it operated from 1887 to 1964, long after the mill closed, and is now a private residence. Students had plenty of playtime in between classes but were expected to do the necessary chores required to keep the school warm in the wintertime and to keep their water source, a spring, cleaned of debris. The schoolhouse, independently owned, made an equally ideal playing field for community events such as picnics and an election polling place. The first Congregational Church service for Falk residents was at the school in 1894. Noah Falk donated all the building materials, and the official church was erected behind the school in 1897. (Both, courtesy of Dick and Patti Bermuda.)

William "Bill" Fleckenstein grew up around the logging industry. His father, Joseph, left Iowa and came out to California working various jobs and, in 1900, as a woodsman in Hydesville. Bill worked as a laborer, and by 1913, he worked his way up to a woodsman and was living in Falk. He met Ida Fleming, whose father, Martin, worked as a steam donkey operator. They had two children—Ernie in 1916, and Dorothy in 1918. There was no official doctor in Falk during this time, so when Ida came running into the Falk general store in labor, Vern Olsen, the town bookkeeper, grabbed the only vehicle in Falk and rushed her to the hospital in Eureka. In 1922, Ida contracted tuberculosis, and Bill was soon widowed with two small children. Ida's sisters and younger brother moved in with Bill to help raise Dorothy and Ernie. In times of necessity, especially during the Depression, many families moved in with each other to save on resources and help with the many required chores and raising of children. (Courtesy of the Callihan family.)

Bill Fleckenstein, the lead chopping boss, and Ida (below) had a home large enough so that Ida's brother Martin could live there and pay board, which was a common occurrence. The children, Dorothy Fleckenstein and her brother Ernie, shared the same household, and it was a busy active place. Ernie was born in 1916 and Dorothy in 1918. Shortly after the birth of her children, Ida contracted tuberculosis, also known as consumption. During this period of time, 110,000 people died each year from this disease; a cure was not discovered until the 1950s. She succumbed to the disease in 1922 and died at 24 years of age. Their aunt Helen Fleming, Ida's sister (at left, between the children), moved in with the young children and raised them until they were of age and able to care for themselves. (Both, courtesy of the Callihan family.)

Dorothy Corvi, daughter of Bill and Ida, came to Headwaters for a visit in 2008 with her two sons, Fred and Mike. She shared how living in a company town was isolating, and when she was a child, her dad had built a dollhouse in a huge redwood stump, and she played with her dolls in it. Someone set the stump on fire, and Dorothy still remembered it 70 years later. In 2015, a Headwaters docent, Patrick Roskam, peered into a stump near Falk and pulled out a tiny dolly head (pictured). The stories of Falk and the artifacts continue to surface well into the 21st century as long a people are paying much-needed attention. Dorothy had left Falk in 1936 for San Francisco, gone to beauty school, married, had children, and lived the rest of her life in San Francisco. (Both, courtesy of the Callihan family.)

Paul Mazzucchi (pictured with Clara at left) was born in 1908 to Swiss Italian immigrant parents who moved to the Elk River Valley in 1918 to do dairy farming. When Paul was strong enough to go to work, he got a job at the Elk River Lumber Company as a woodsman in 1923. He would walk the mile and a half each morning to catch the 6:00 a.m. train from Falk to Maggie's Camp, where he would eat his breakfast and get lunch. Paul eventually became a rigging operator for several logging companies in his logging career, including Hammond, Carson, and Redwood Lumber Company. He became a full-time dairy farmer in 1940 after his parents retired and he bought the business and land. He was instrumental in telling the story of Falk and the history of the Elk River valley. His interviews are used as primary sources and are a valuable asset to Humboldt County history. (Left, courtesy of Paul Mazzucchi; below, author's collection.)

Born in 1906 to Edward and Lottie Callihan, Shirley, the youngest of four children, spent his entire childhood in Falk. Edward supported the family on a substantial income due to his mill foreman position of 10 years. They had one of the finest houses in Falk, and it was located across the river from the mill. To grow up as a boy in Falk meant that there was fishing, hunting, and playing in the backyard woods. Shirley, pictured at right, spent every free moment fishing and enjoyed this activity throughout his adulthood. The South Fork Elk River during this time was rumored to have had so many salmon that it would be easy pick them up out of the water or walk across their backs. Shirley went on to marry and raise his family outside of Falk. He made a career as a mechanic and worked for the Pickwick Stage company in Eureka. (Both, courtesy of the Callihan family.)

Shirley Callihan and Helen Fleming were Falk neighbors in their formative years. Many Falk children spent so much time together at school, dances, and in social settings that marriage was ultimately the natural progression. It was at one of these Falk dances that they fell in love and started their courtship. Around this time, in the early 1920s, vehicles and vehicle ownership were more common, giving the Falk residents opportunity for more freedom. Couples like Shirley and Helen could spend a few hours in the evening going to a dinner and a movie and be back for curfew. After they were married, Shirley commuted to Eureka for work, which was unheard of in previous years. The luxury of the automobile changed the dynamics of company towns with both entertainment and work opportunities. The couple eventually had three children together—Shirlene, Elve Mae, and Dale Callihan. They lived a short time in Falk before setting out for the big city of Eureka in the late 1930s. (Both, courtesy of the Callihan family.)

Although Falk was only eight miles from Eureka, it was incredibly remote. The only means of transportation in the early 1900s was walking or by horse and buggy. In the 1920s, the Nellist Brothers Stage Line became a viable option to get major supplies to and from Eureka. Shirley (above) and Art (right) Callihan, brothers who grew up in Falk, became stage drivers and mechanics for two different companies, the Pickwick Stage and the Nellist Stage line. Shirley worked for the Pickwick, which operated primarily on the West Coast. The Nellist Stage, driven by Art Callihan, transported Falk students to junior high and high school. The stage came to Falk twice daily and delivered goods including groceries, shoes, watches, and even luxury items. Before the automobile became commonplace, the residents relied on the stage to expand their choices in an otherwise isolated place. (Both, courtesy of the Callihan family.)

Handcars, or velocipedes, were used as a transportation method for the Falk community. This was an effective way to get to Eureka without having to ride a horse or walk. In 1887, Noah Falk found himself late one night at the mill missing the last train into Eureka. He jumped on a velocipede and headed in the direction of Eureka. He was determined to chase his train down. However, one of the crossbars came loose, he went to adjust it and flew off the trestle into the Elk River. Uninjured, he found his way to Eureka unscathed. Aside from mishaps like these, handcars were a viable option for transportation. (Left, courtesy of the Callihan family; below, courtesy of the Boyle Collection, Humboldt State University Library.)

Without the Bucksport & Elk River Railroad, the town of Falk and the Elk River Mill and Lumber Company would have been nonexistent. The residents primarily used the stagecoach for their own transportation, and the rail line was designated solely for the movement of lumber and commerce to Bucksport or Falk. The line was also used as an avenue for employment and for some of the valley residents to make extra money. Many of the upper valley inhabitants supplied the cookhouse with cords of wood, milk, potatoes, and vegetables. They would stage their products by the railroad tracks, and the engineer would stop the train and load the produce up to the cookhouse. Paul Mazzucchi made rent money when he grew potatoes for the cookhouse and made $5 for a 100-pound sack. A Falk resident poses on the tracks leading out of town, once a source of pride—in contrast to the decayed trestle depicted below in the 1970s. (Right, courtesy of the Callihan family; below, courtesy of Leonard McLaughlin.)

With the advent of the automobile, the company town dynamics changed. No longer did people have to live in town; they could commute and live in neighboring communities such as Eureka. Clarence Weatherby (pictured at left) later took over as the railroad superintendent of the Bucksport & Elk River Railroad after James Wrigley left. Clarence's brother James took over the job as the railroad engineer. Although there was a railroad going from Falk to Bucksport, seats for passengers were limited. A reprieve, and a new freedom for Falk residents, came with the invention of the automobile. (Above, courtesy of Jon Humboldt Gates; left, courtesy of the Callihan family.)

Company towns were small snapshots of the larger American landscape of hard work, community, and the blending of families. Zella Callihan married Clyde Moore in 1920. She had grown up in Falk and was the daughter of Elk River Mill lead sawyer Ed Callihan. Clyde Moore's father, Lorenzo Moore, migrated in the 1880s from New Brunswick, Canada, where he had gained the skills necessary for the logging industry. Lorenzo settled in Humboldt County and eventually became a night watchman in Loleta in 1920. His son Clyde found work in Falk as a woodsman and, eventually, in the railroad industry. He and Zella married, and by 1930, the extended family was all living together in Falk, taking care of Lorenzo, who passed in 1936, one year before the mill closed for good. (Both, courtesy of the Callihan family.)

The only day off in the busy town of Falk was Sunday. This was a special time during the week, after church, where the families would get together and have a picnic at the Jones Prairie School location. It was also a place to bring a favorite dish, as well as the produce from the garden, to share with others. It was a chance to wear Sunday best and pose for photographs, as seen here. In the wintertime when it was raining, residents had their picnics on a boxcar with a roof over them to stay dry. Meeting like this forged bonds with the children and adults who shared a great meal and played a fun game, all the while conversing about the past week and future plans. (Both, courtesy of the Callihan family.)

Throughout the years, the stories about growing up and working in Falk were always shared with a smile and wistful expressions across the proud faces of the storytellers. They shared about the time they went to dances, sat on the porch knitting, gossiping about the latest crush, shared meals, fishing, wading through the Elk River on warm summer days, and, above all, walking through the land of sawdust. In such a tight river valley, the ample sawdust from the mill blew in the air they breathed, covered the paths they walked on, and rested on every surface they encountered. A few years back, some steps were built down to the river near "downtown Falk, California," and in place of soil was sawdust dating back over 100 years. (Both, courtesy of the Callihan family.)

As the Great Depression approached in the 1920s, job positions transitioned from the early prosperous Falk to a waning lumber town. The prior state-of-the-art steam machinery and equipment was in need of critical upgrades. The old-growth redwood trees once close to the mill were now out in the far reaches of the Elk River watershed. The thriving town of Falk was losing ground. By 1930, the mill closed, and some residents remained. Ed Quinn took over the responsibility of running the general store and continued to live in the sparsely populated town with his wife, Elve (Fleming), until the closure in 1937. (Both, courtesy of the Callihan family.)

Paul Mazzucchi, a local Elk River resident and former Falk employee, recounted to the author in his 2002 interview why the Elk River Mill closed: "I know they were having a hard time there because it had been going on a week or two and we came home one night on the train. We was logging up the river, nice timber, the number one. We got up near the roundhouse and we was walking down. There was Bill Fleckenstein, he was the foreman. 'Paul,' he said, 'look at the logs down there, number one logs.' He just saw them trees, 'now they're going to butcher them up. The mill is cockeyed and they're going to make wedges out of them. One end was big and the other skinny. The mill was jackknifed. All we doing up there is logging and they waste the timber and all that.' That's why they shut down. I could see it from here when the train would go by with lumber. More lumber came back than what they shipped down there. They blamed it on the depression. We were getting the logs in the pond but they were butchering them in the mill. When you throw lamb into potatoes, there's no money coming in. They borrowed a million dollars, they done away with the donkeys and bought Caterpillars, two of them. We were getting the logs in but there was no money coming in." (Courtesy of the Callihan family.)

Life in Noah Falk's company town relied upon an intriguing set of beneficial characteristics, of which adaptability and resiliency were key. When disaster struck, the resilient nature of the town's men and women came to the fore. Having built their workplaces and homes from locally derived materials, community members drew again from local materials and the skills of company town employees to rebuild following fires or floods. A fire wiped out the Elk River Mill, yet Falk, his family, and his employees rebuilt, resurrecting the company town's primary place of employment with trees downed, milled, and erected on the spot. Adaptability helped company towns to succeed and grow. Nonetheless, company towns including Falk were dependent on overall market conditions for continued success. As markets shifted, company towns were tested. For Falk, an unprecedented downturn in the timber market in the early 1930s marked the beginning of the end for the Elk River Mill. Unable to weather this downturn, Falk fell victim to the boom-and-bust nature of existence common to resource extraction industries around the globe. (Courtesy of the Callihan family.)

Three

Ghost Town

Noah Falk's claim was still thick with vast stands of redwoods—a valuable commodity. In the early 1950s, the entire property was purchased by Cheney Lumber of Washington State. A young, adventurous forester named Riggs Johnston was hired by Cheney in 1954 to manage these timber holdings, which surrounded the town of Falk. At that time, the town's buildings were fairly intact, a locomotive still sat on the rails, and a few year-round residents still called Falk home. In the 1950s, 1960s, and into the late 1970s, Riggs would take his family and Boy Scout troops on many outings to the ghost town to hike, explore, or camp. Memories of those outings include picking fruit off the abundant apple trees, walking across the railroad trestles, perusing the general store with mail still in the slots and shoes still on the shelves, water still dripping from faucets, the bowed and decaying floors, huge rats' nests in a lot of the structures, and the timber skeleton of the Falk mill. There are also fond memories of Charlie and Loleta Webb, the local couple hired by Cheney for security. It was clear on these outings that Riggs adored this town and its history, and he was caught in a position of conflict. In the 1960s and 1970s, there were not the preservation efforts that there are today. These decaying structures would prove to be an attractive nuisance to so many, and much to Riggs's dismay, "bottle hunters" would trespass regularly to loot what they could. Time spent policing the town proved to be too much, and Mother Nature had also taken her toll, making most of the buildings unsafe. The passing of Charlie Webb and the insurance liability of the unsound structures proved to be the town's death knell. The decision was made to donate some of the structures to the fire department for training purposes and knock down the rest. Years later, when Riggs was reminiscing about the company's decision to raze the mill, sadness filled his eyes.

Shortly after the final mill closure in 1937 due to high operating costs and limited sales, the structures that once thrived with life sat vacant as if time stood still. The company had hoped that someday it would reopen the mill. Tools and machinery like steam donkeys were left in the woods, equipment was left in the mill, items were left on the store shelves, and furniture was left in the houses. Stepping into one of these buildings in the 1940s was like stepping back in time. Dishes were left in the cookhouse, and an industrial stove that cooked pies, bread, baked goods, breakfast, lunch, and dinner looked fairly operational, as seen here. Rich Wills, a local Eureka resident, shared in his story about the ghost town: "Moving through the town area we went up the canyon to what must have been the residential area, Quite a few homes were still standing at that time but mostly had no furniture or window coverings. Wallpaper—woodstoves—and I remember a couple had wood-burning kitchen ranges." (Courtesy of the Clarke Historical Museum.)

Constructed in the late 1800s with Elk River Milled old-growth redwood, this cookhouse was fading into the landscape by the 1960s. The crushing vegetation and rot were virtually collapsing the buildings. Trespassers faced a real danger when entering these structures, such as the cookhouse (pictured), when they insisted on going to the second floor and risked falling through the rotten structures. They also faced being caught by Riggs Johnston or Charlie Webb. Rich Wills, a local teenager who was trespassing and was almost caught by Johnston or Webb, recalls, "When we decided to make our way back to the main road we heard someone yell at us and we ran like the dickens. I kept waiting for the pack of dogs to overtake us but we never did see any dogs." By 1979, the company saw these buildings as a liability, and all remaining structures were burned and bulldozed. (Above, courtesy of Sy Beatte; right, courtesy of Jon Humboldt Gates.)

In no time at all, buildings such as the saw filer's shop (pictured) took on an empty feeling in contrast to the traditional bustling team of men sharpening bandsaws on a daily basis. These buildings and the equipment left behind created a magnet for bottle hunters and high school students looking for fun; as Rich Wills described, "I was fortunate enough to be invited in what turns out to be an adventure back in the summer of 1967 as I remember it. We had just graduated from Eureka High School in June. A few friends invited me along on an excursion out to the town of Falk, or what was left of it. I recall we had to park on the main road and hike through the woods because there was a caretaker of the town at that time who would run us off if he encountered us. Rumor was that he had a pack of dogs and he would set them on you if he did see you. Also had a shotgun loaded with rock salt. Probably just rumors but they sounded true at the time." (Courtesy of the Clarke Historical Museum.)

Loleta Webb, formerly Loleta Barnes, lived in Falk from the early 1900s to 1975. She and her second husband, Charlie, were the caretakers of Falk from 1950s to the 1970s. Their house was technically outside the boundaries of Falk; in order to see the ghost town, one had to pass their home. Hired by Elk River Lumber Company, they had the job to keep people from trespassing into the old buildings. Charlie was known to patrol with a shotgun full of rock salt, tap people on the shoulder, and asked them to drop everything. The more experienced trespasser followed the railroad grade that bypassed Charlie and Loleta's house. Although the community of Falk had all but disappeared, their presence was an essential part of its history. Presently, there are remnants of their house a quarter mile from the Elk River Trailhead. (Courtesy of Sy Beatte.)

The yew trees pictured in front of the Webb home can still be seen underneath a new canopy of redwood trees. Even though the yew trees were introduced, they were a sign of community life, much like the fruit trees that still blossom every year. The house was razed shortly after Loleta's death in 1975, just short of her 88th birthday. The outlines of her water tank perched on top of a nearby massive redwood stump, the foundation of her porches, a scattering of pipes that still protrude from the ground, and a kitchen sink are all evidence of the progression of modern conveniences of the 1960s enjoyed by the Webb family. Today, approximately 800 schoolchildren visit this site, and each enjoys a living history presentation and the chance to explore the vestiges of a house vanishing under a new generation of redwood trees planted in the 1970s and early 1980s under the supervision of lead forester Riggs Johnston. (Courtesy of the Callihan family.)

Loleta and Charlie Webb virtually saw the town of Falk disappear before their eyes. As caretakers for more than two decades, they saw the buildings slowly fade into the landscape. They were able to grow their own vegetables and get their water from the Elk River and a catchment system on a stump next to their house. Charlie (below) apparently had a cigar in his mouth always, whether he was holding a deer or swimming the backstroke in the Elk River. Originally from England, he was able to come to America after winning enough money in the Irish Sweepstakes. He and Loleta married in 1949 and continued to reside in the area just west of Falk until their deaths, Charlie's in 1972 and Loleta's in 1975. (Right, courtesy of Sy Beatte; below, courtesy of the Callihan family.)

Richard Lisignoli, who was nicknamed "Dick" or "Garibaldi" (for the former Italian general), was the last official Falk resident who lived in the immediate vicinity of the town. At the time of this photograph, he is standing on the porch of his cabin, which was located immediately adjacent to the Elk River Mill cookhouse in downtown Falk, California. This was taken before the cabin burned in 1961. At this time, there were several residents outside the Falk town proper. Stillman and Stedman Fryers lived in the Fryers home across Elk River, about a half mile from the present-day Headwaters Reserve parking lot. The Fryer twins were working in Eureka but still taking advantage of the low rent in Falk, which was only $5 a month. Also during this time, Loleta and Charlie Webb were the caretakers of the decaying buildings of Falk, and they lived a quarter mile from the parking lot. (Courtesy of Jon Humboldt Gates.)

Dick worked as a sawyer for Elk River Lumber since 1907 and had originally come from Italy. In a 1923 wage sheet, he is listed as a crosscut sawyer at the age of 47; this was a difficult job for a young man, much less someone in his 40s. Dick had saved all his money in Italian bank accounts so when he retired he would come back to his home country and live out the rest of his days. Around the time of the mill closure 1937, he wired for his money and received a telegram back saying that his life savings were confiscated and now state property. With few options and an aging body, he remained in a Falk cabin with low rent. Taking full advantages of the resources around him, he fished, hunted, kept a garden, and lived an overall frugal life in his Falk cabin until it burned down in 1961. He then moved to Fortuna and died in 1963 at the age of 87. (Courtesy of Jon Humboldt Gates.)

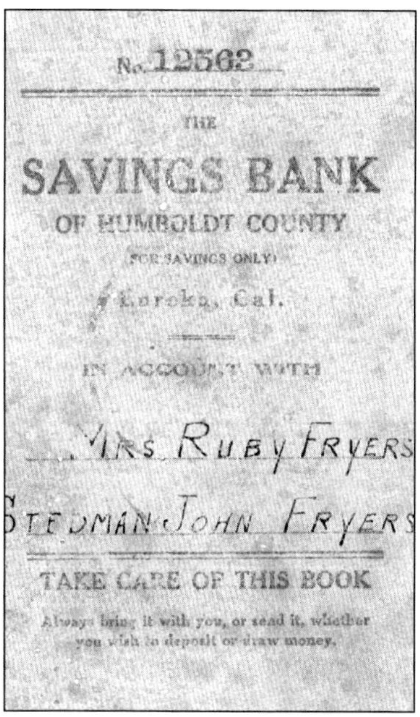

John and Ruby Fryers lived across a footbridge about a half mile from the Falk mill. John was a mill worker in 1914 when his twin sons, Stillman and Stedman, were born. Ruby was kind enough to teach all the young "bucks" in town how to dance before the Saturday night dances and had to endure the boys walking all over her feet. She may possibly be credited for the enduring relationships that were formed after those evening dances. She also taught piano lessons for a whopping 50¢. Her twin boys were infamous in several oral interviews and kindly referred to as bachelors who lived on and off at their Falk home until they died in 1975 and 1981. Both of them had careers as woodsman throughout Humboldt County and were drafted in World War II. The house they lived in was bulldozed after 1981 when Stedman died; he was the last anchor resident of Falk. (Above, courtesy of Sy Beatte; left, courtesy of Robert Ehrhardt.)

Stedman John Fryers's bank book (pictured) was recovered in the 1980s by an Elk River resident, Bob Ehrhardt, just before the Fryers' house was razed. The bank account was opened in October 1915, when Stedman was one year old. Ruby Fryers, Stedman's mother, opened the account at the Bank of Eureka, which is presently the Clarke Historical Museum. Coincidentally, one of the bank's founding directors was Noah Falk. The first deposit was for $1, roughly equivalent to a $25 deposit in 2018. Stedman liquidated his account in 1938 at the age of 24. Between these years, the account grew to hold just over $112 in 1932, the equivalent of $2,000 now. Future visitors to Headwaters Reserve will get the opportunity to walk by the Fryers' homesite in 2019 when the new trail is completed. (Right, courtesy of Jon Humboldt Gates; below, courtesy of Robert Ehrhardt.)

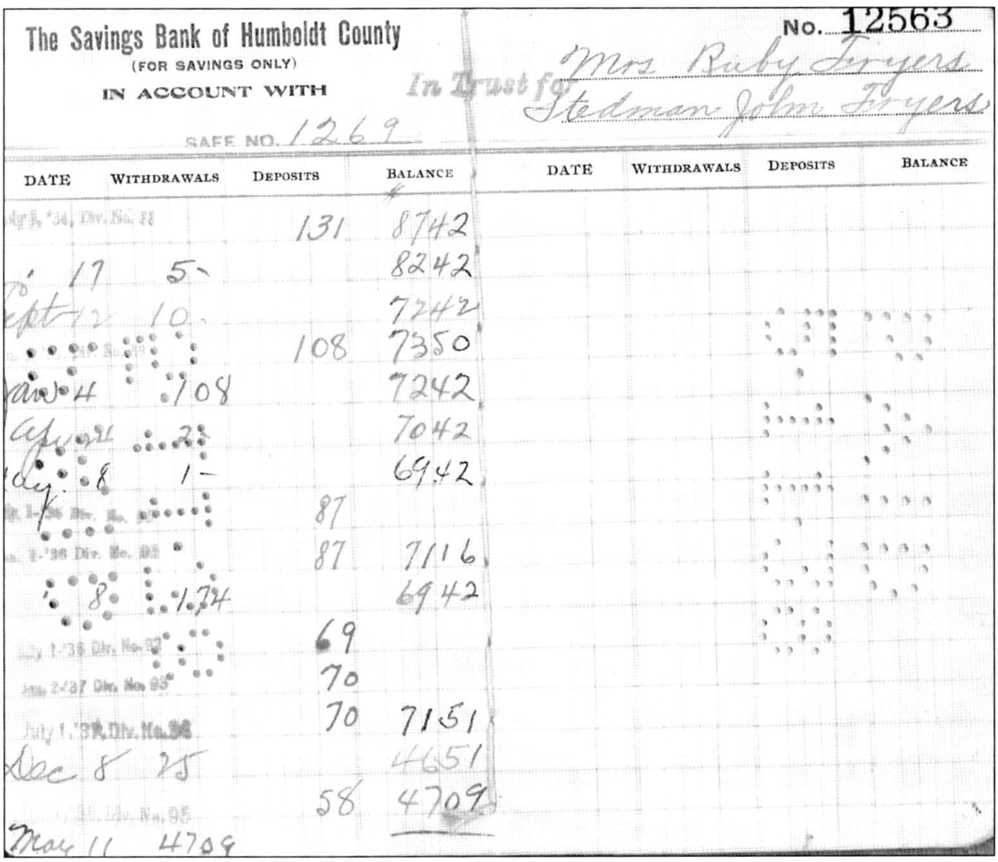

Memories from the 1960s and 1970s of Falk made quite the impression on local teenagers who trespassed to see the ghost town. Michael Congdon from Ferndale wrote in the *Eureka Times Standard* about his memory of Falk in the 1960s: "All us kids would hike in with a compass from the back door side of Falk and stay the nights under an old over grown apple orchard during the mid and late 1960's. I can still see stretched out in my mind's eye the carpet of glow worms we slept with, gleaming like a million stars. I loved that place and to this day I always have found myself thinking about those times spent out there as a young kid. There was a lot of stuff in the town back then, but we took nothing away with us because we thought it would be bad luck to do so. Kids in a ghost town, our ghost town has a way of breeding respect. And now I learn that they came in and bulldozed it all. What a shame. There was a sort of back street with a row of small cabins that had a line of rabbit hutches. I could peel the layers of newspaper back that lined the floors and sit reading half day long." (Courtesy of Humboldt County Historical Society.)

The once innovative bandsaw first used in the Elk River Mill is now part of the forest floor. The town of Falk, the mill itself, the tools, bull donkey, two locomotives, and anything that could be salvaged was loaded onto a train and taken downriver. The final elements to be dismantled during this period included the logging railroad itself. However, present-day visitors to Headwaters Reserve have come across artifacts, like the sail of a 1920s Morse windmill, that are left in place for all visitors to enjoy, present and future. The Antiquities Act of 1906 established the first general legal protection of cultural and natural resources in the United States. The townsite and surrounding area in the Headwaters Forest are protected by this act and the 1979 Archaeological Resources Protection Act. If one finds an artifact, it must be left in place. Signs along the trail inform visitors that artifacts are protected for the enjoyment of all. (Courtesy of Humboldt County Historical Society.)

In the 1960s, Jon Humboldt Gates came out to the ghost town of Falk with permission from Charlie Webb. He went into the Falk store and found shoes on the shelves waiting for repair, dishes in the cookhouse, and bandsaws in the Elk River Mill. Inspiration from his experience and the gracious stories told to him by past residents led to his initial College of the Redwoods Falk research paper. He eventually learned that his great-grandfather and grandfather lived in the township of Falk. Although he had family ties to the community of Falk, Gates himself was from the 1960s generation and most of the people he was going to interview were born in the early 1900s. Not surprisingly, when Gates knocked on their doors, he was met initially with some skepticism, but they warmed to him and the relationships that developed led to countless hours of reminiscing about life and Falk in the Elk River valley. He recorded the interviews on tape, but the tapes were later stolen. Fortunately, Gates had transcribed most of the tapes and was able to use them for *Falk's Claim: The Life and Death of a Redwood Lumber Town*, his history of the town and its residents. (Courtesy of Jon Humboldt Gates.)

When Jon Humboldt Gates entered the town of Falk, he discovered items such as these dishes left in the cookhouse that had once held so much activity. The year was 1968, and most of Falk's residents had been gone nearly 30 years. As luck would have it, some of the Falk residents were still in the surrounding community, allowing Gates to reach out and follow up with semiformal interviews. From these interviews came countless stories, alive in the storyteller and clearly worthy of capturing. Gates interviewed former residents, including Chester "Ted" Barnes, Vernon Olson, Irving Wrigley, Loleta Webb, and many others, and held a Falk reunion in 1981. From the time he started his research paper to the final *Falk's Claim* book, it took him 10 years. He dedicates the book to his mother, who typed the draft three times on a manual typewriter. (Courtesy of Jon Humboldt Gates.)

By 1963, the land that was once Falk was owned by Sierra Pacific. During this time, the mill had been closed approximately 30 years. With heavy yearly rainfall, the buildings deteriorated at a faster rate compared to ghost towns like Bodie, California. The vegetation pushed through the cracks and crevices, and nature reclaimed the vacant buildings. Riggs Johnston, the lead forester, spent much of his time protecting the decaying buildings from trespassers and pillagers. His children Chris and Patti Johnston are pictured at left near the once thriving town. (Both, courtesy of Pete Johnston.)

The Elk River Mill and Lumber Company closed its doors in 1937 due to high operating costs. For 40 years, equipment, lumber, automobiles, working machines, and furniture remained in the adjacent houses and buildings. This wood cart was used to dump sawdust away from the Falk Mill. In 1961, when this photograph was taken, it had been sitting in storage for over 20 years. (Courtesy of Jon Humboldt Gates.)

The Elk River Mill and Lumber Co. No. 3 was new to Falk around 1922, the Heisler transported logs to the Elk River Mill. After the town's demise in the early 1950s, it was partially scrapped and sold to Mutual Plywood in Fairhaven. Mutual Plywood converted it to a diesel engine using a Murphy power unit. It eventually found its way to the Samoa roundhouse, where it sits today. (Courtesy of Jon Humboldt Gates.)

The Falk roundhouse, also referred to the train barn, was located across the Elk River and upstream from Falk about a half mile. According to a local newspaper, it was built in March 1900. A train mechanic and his small crew's full-time job was to keep the three locomotives in operation and use this facility for train repair and maintenance. Jon Gates closes his book 1983 book *Falk's Claim* with these final poignant sentences: "Their existence has been subject to the rhythm of nature which evolves, thrives, and in time, grows extinct, through a perpetual cycle that advances throughout creation. With this thought in mind, I picked myself up off the dusty floor of the train barn, and with a last look around at the emerging forest, left what had once been Falk's Claim." (Both, courtesy of Jon Humboldt Gates.)

In 2008, this historic roundhouse building site was excavated by Humboldt State University cultural resource staff. The archaeologists discovered evidence of another building that predated the 1900 structure and may have been a blacksmith shop built around a similar time frame as the Elk River Mill. After the 2008 excavation, the roundhouse was salvaged and moved piece by piece to its current location on the Elk River Trail. The barn was built out of old-growth redwood, and due to its resistance to rot, 20 percent of the building was recovered and reconstructed with historic restoration carpenters/teachers. Carpenters, teachers, and students from College of the Redwoods, Colorado Mountain College, Mountain Heritage Associates, and the US Forest Service Enterprise Program collaborated on the project. It now serves as the Headwaters Education Center and is open in the summer months on Sundays and for special events. It is located a half mile from the Elk River parking lot. (Both, courtesy of Jon Humboldt Gates.)

Just after it was constructed in 1890, the mill ignited with a single spark and burned rapidly due to all the fine sawdust that covered the floor and equipment. It was rebuilt shortly thereafter, and strict measures were developed to keep fire hazards at a minimum. By the 1900s, the Elk River lumber mill was producing 40,000 board feet per day. The by-product of all this success was the copious amount of sawdust. Before the automobile, all this sawdust was moved by hand to large piles adjacent to the river (pictured). Edward Olson's official job title was "saw dust man," and he made $60 dollars a month plus board in 1923. (Both, courtesy of Jon Humboldt Gates.)

By the latter 1920s, a Model T dump truck did most of the work. Later, after the mill closure in 1937, the equipment was left in a shed and never operated again. The Model T slowly deteriorated under the shack until that collapsed on it too. The shed and the Model T are now part of the forest floor; the remnants of the chassis are still visible. A memory of the Model T was shared by Michael Congdon, who visited Falk in the 1960s: "There was an old Model T car parked inside a garage. I would sit in it and I would yell and wave pretending to drive down Main Street on a warm summer's day." (Above, courtesy of Sy Beatte; below, courtesy of Jon Humboldt Gates.)

By the 1970s, the forest was reclaiming the town into forest ground. The formerly maintained and thriving bustling mill town was disappearing. The houses that were once painted took on the look of surrounding woods, with their boards baked by the sun. White picket fences that were a sense of pride sank toward the earth as the weight of plants pulled them down. The dance hall sat vacant, with two pianos never to be played again, and all that remained of the dance floor where Ruby Fryer's dance lessons took place were memories. One got a sense of a timelessness mixed with recollections of a different era and a door closing forever. (Both, courtesy of Humboldt County Historical Society.)

These bachelor cabins survived years of non-use following the departure of the men who once inhabited them for 50 years, 1886–1937. Built out of old-growth redwood and post and pier foundation, the cabin had space for a woodstove and two beds. Onlookers who visited these structures during the ghost town period commented on how sturdy they were built in light of the encroaching blackberry vines and ivy. Also mentioned was that the interior walls were covered with newspapers dating back to the late 1800s, covering up the drafty holes. (Both, courtesy of Jon Humboldt Gates.)

The men that found their home here may have traveled from countries like Finland, Sweden, and Ireland. During the Depression, the lumber mills were a viable workplace still employing men. Paul Mazzucchi, an Elk River resident, remembers seeing men walking past his house with a bedroll and a backpack, asking for directions to Falk. (Both, courtesy of Jon Humboldt Gates.)

Between 1900 and 1930, Elk River Mill superintendent Charles Falk and his wife, Blanche, lived in a two-story house overlooking the Elk River Mill. Company town structures and their placement design was strategic and intentional. The owner's house was almost always on top of a hill overlooking the workforce. The mill cookhouse and the bachelor cabins all surrounded the Elk River Mill. The remainder of the town was spread throughout the valley with space limitations dictated by the Elk River, the narrow canyons and a lack of earth-moving equipment. This made for a tight fit for all the necessary housing for 200 residents. (Above, courtesy of Sy Beatte; below, courtesy of Jon Humboldt Gates.)

Leonard McLaughlin, an Elk River resident, recalls in the Summer 2009 issue of the *Humboldt Historian* that he went out to Falk after a winter's storm in 1977. McLaughlin entered the superintendent's house and found that there was still china in the cupboards and kitchen utensils in the drawers but that climbing the stairs was daunting because it was not the soundest part of the house. On the second floor, he had a view of the mill and knew exactly why this house was built there. (Both, courtesy of Leonard McLaughlin.)

The Bucksport & Elk River Railroad train trestles were all but falling apart by 1977. These strong structures had supported the weight of trains and the lumber that left Elk River Mill for 50 years. What used to be a path to the Falk ghost town around Charlie Webb's house became hazardous for any kind of hike. The train tracks were pulled in 1949 as the last train had left Falk in 1937. The lifeblood of commerce and transportation was now falling into the Elk River. The trestles ran parallel up the watershed approximately three miles from the present-day Elk River parking lot. In the winter months, with the lack of vegetation, visitors can see remnants of the trestles with the sturdy posts the size of telephone poles nailed to the nearby redwood trees. It is hard to believe that in this now-lush forest trains were running up to Maggie's cookhouse, where men would eat a huge breakfast before a day of logging operations. (Above, courtesy of Leonard McLaughlin; right, courtesy of BLM.)

These former cabins lodged a chief part of the Falk workforce. On a hand-drawn map of Falk, there is a bachelor cabin neighborhood across the river called "China town." Although the bachelors came from across the country and the world, they were not Chinese. Logging camps were generally a homogeneous European population. Room and board cost $5, and the monthly wage in the woods averaged $100 a month, while the mill workers earned an average of $60 per month in 1921. Paul Mazzucchi, a "riggin man," remembered what a typical day of work was: "Up at 6 o'clock in the morning, eat, walk to work, jump on the train and come home and some of the guys living around there, single fellows, they have cabins around China town and they play cards." He also reminisced about sharing a cabin with a more studious cabinmate who studied law books nightly under a kerosene light and eventually became a judge. (Above, courtesy of Jon Humboldt Gates; below, courtesy of the Clarke Historical Museum.)

This Craftsman-style house was one of the last houses built in Falk around 1920. Vernon Olsen, the Falk bookkeeper, lived here with his wife, Ruby, and their two sons, Lewis and Lynwood. The house sat across the Elk River, and to access it, the family had to take a footbridge from what is now the Elk River Trail. The house was one of the only Falk homes with a flushing toilet and modern conveniences, such as a water heater for the washroom. The nearest neighbors were the Haleys, and they shared a sidewalk that led to the footbridge. The Olsens lived in this house until 1932, and Vernon got a job in Santa Rosa, California. Shortly thereafter, the blacksmith John Miller moved here with his wife, Ruby, and their children, Wayne and Evande Miller. The house collapsed in 2013 and was the last standing intact Falk home. (Both, courtesy of BLM.)

Bibliography

Allen, James B. *The Company Town in the American West*. Norman: University of Oklahoma Press, 1966.

Carlson, Linda. *Company Towns of the Pacific Northwest*. Seattle: University of Washington Press, 2003.

Clark, Julie D. *Company Towns in America: 1880 to 1930*. Arcata, CA: Humboldt State University, 2006.

Crawford, Margaret. *Building the Workingman's Paradise: The Design of American Company Towns*. London: Verso, 1995.

Gates, Jon Humboldt. *Falk's Claim: The Life and Death of a Redwood Lumber Town*. Seattle: Moonstone Publishing, 1983.

About the Historical Society

Since 1947, the historical society has worked to preserve original historical materials related to Northern California history. Membership, which originally stood at 117, now numbers nearly 3,000, making us one of the largest county historical organizations in the country. Members receive the *Humboldt Historian*, our quarterly magazine, voting privileges, and discounts in our bookstore. Membership is open to everyone. Our purpose is to further an understanding and appreciation of all peoples, places, events, and activities of Humboldt County and related areas. Toward this end, we focus on acquiring, preserving, interpreting, creating, and sharing historical information and by educating and assisting others to do the same. To further these goals, the Humboldt County Historical Society publishes an outstanding quarterly magazine, the *Humboldt Historian*; maintains a thorough reference library; prints historical books; records and transcribes oral histories of knowledgeable residents; archives valuable collections, and makes the information in those collections available to researchers. All royalties from the publication of this book will be donated to the society.